# Luminescence - Basic Concepts And Emerging New Applications

*Edited by Ahmed M. Maghraby*

Published in London, United Kingdom

Luminescence - Basic Concepts And Emerging New Applications
http://dx.doi.org/10.5772/intechopen.1002556
Edited by Ahmed M. Maghraby

Contributors
Alexander I. Ignatev, Amir Hossein Farahani, Amir Moussaoui, Daihao Li, Dalia Chávez-García, Dmitry V. Bulyga, Helena Cristina Vasconcelos, Mario Guzman, Mingqing Wang, Natalia K. Kuzmenko, Nikolay V. Nikonorov, Reza Zarei Moghadam, Robert Moss, Sergey K. Evstropiev, Stephen Kearney, Tianming Sun, Yohichi Kohzuki



First published in London, United Kingdom, 2024 by IntechOpen
IntechOpen is the global imprint of INTECHOPEN LIMITED, registered in England and Wales, registration number: 11086078, 167-169 Great Portland Street, London, W1W 5PF, United Kingdom

British Library Cataloguing-in-Publication Data
A catalogue record for this book is available from the British Library

Additional hard and PDF copies can be obtained from orders@intechopen.com

Luminescence - Basic Concepts And Emerging New Applications
Edited by Ahmed M. Maghraby
p. cm.
Print ISBN 978-0-85466-893-9
Online ISBN 978-0-85466-892-2
eBook (PDF) ISBN 978-0-85466-894-6

For EU product safety concerns:
IN TECH d.o.o., Prolaz Marije Krucifikse Kozulić 3, 51000 Rijeka, Croatia,
info@intechopen.com or visit our website at intechopen.com.

# Meet the editor

Ahmed M. Maghraby graduated from the Faculty of Science, Cairo University, Egypt, in 1995, and started his career as a scientist at the National Institute of Standards (NIS), Egypt. He received his MSc and Ph.D. from Cairo University in 1999 and 2003, respectively. His Ph.D. work focused on the calibration and characterization of organic materials for use in radiation dosimetry using electron paramagnetic resonance (EPR). In 1999, Dr. Maghraby worked as a guest researcher at the National Institute of Standards and Technology (NIST), Maryland, USA, and as a visiting scientist at Dartmouth College's EPR Center for the Study of Viable Systems, New Hampshire, USA, in 2008. In 2006, he worked as a guest researcher for the Physikalisch-Technische Bundesanstalt (PTB), Germany's national metrology institute. Dr. Maghraby has led and participated in several international research projects on radiation measurements. His research interest is ionizing radiation metrology, especially ionization chambers, EPR, and thermoluminescence dosimetry (TLD). He is a reviewer for several peer-reviewed journals, has supervised several master's and Ph.D. theses, and was elected as the vice president of the National Committee of Physics (NPC) in Egypt.

# Contents

# Preface

Although luminescence has been observed since ancient times and throughout early human civilizations, the term "luminescence" was introduced by the German physicist Eihardt Wiedemann in 1888. Luminescence refers to the emission of light quanta and has many applications in modern life. It can be classified into several types based on the method of production:

- Chemoluminescence: Observed during some chemical reactions
- Bioluminescence: A specific type of chemoluminescence found in certain biological systems
- Electroluminescence: Related to electrical interactions, such as those in light-emitting diodes (LEDs)
- Cathodoluminescence: Produced when certain materials are bombarded with beams of charged particles, as seen in electron microscopes
- Mechanoluminescence: Generated in some materials subjected to mechanical actions like ripping, crushing, or stretching
- Thermoluminescence: The re-emission of absorbed energy upon heating, while cryo-luminescence occurs during cooling of the luminescent materials
- Photoluminescence: Occurs when materials are exposed to light and can be further divided into fluorescence (instantaneous light emission) and phosphorescence (delayed light re-emission)

The mechanisms of luminescence production involve three stages:

1. Excitation: Energy is transferred from the energy source to electrons, exciting them to higher states
2. Relaxation: Excited electrons return to their ground state
3. Emission: Photons are emitted during the transition from higher to lower energy states

The characteristics of emitted photons can vary according to several factors such as the type of the luminescent material, type and concentration of defects and impurities, temperature, and external influencing fields.

Luminescence has numerous applications, including sensors, security systems, biological imaging, displays, and lighting systems, among others. This book covers various topics related to the theory, production, and real-life applications of luminescence and luminescent materials.

**Ahmed M. Maghraby**
National Institute of Standards,
Giza, Egypt

Chapter 1

# Luminescence of X-ray-Irradiated Single Crystals under the Influence of Compression

*Yohichi Kohzuki*

## Abstract

Analyzing the data on bleaching with the *F*-light at the excitation bandpass 5 or 20 nm, it was found that the absorption spectrum of $KCl:Eu^{2+}$ (0.1 mol%) single crystals exposed to X-ray has a peak due to $F_z$-centre within the wavelength 20 nm near *F*-centre peak. Deforming the X-ray-irradiated $KCl:Eu^{2+}$ (0.02 mol%) single crystal by the compression at 100 $\mu$m min$^{-1}$, a new peak is observed around 330 K on the thermoluminescence (TL) glow curve. Its color centre has a new energy band near *F*-band, which is based on the *F*- and thermal-bleaching effects of the compressed crystal.

**Keywords:** thermoluminescence, X-ray irradiation, color centre, *F*-bleach, thermal-bleach

## 1. Introduction

$KCl:Eu^{2+}$ is one of the efficient photostimulated luminescence (PSL) phosphors and has an excellent sensitivity as a X-ray imaging sensor utilizing an optically stimulated luminescence (OSL) phenomenon in comparison with pure KCl [1]. It is well known that the rare-earth dopants, $Eu^{2+}$ ions, are fluorescent and cause strong luminescence. $KCl:Eu^{2+}$ exhibits excellent storage performance and is reusable material for radiation therapy dosimetry [2, 3].

Co-doped KCl single crystal (i.e. $KCl:Eu^{2+}$, $Ce^{3+}$) acts as a potential thermoluminescence (TL) and OSL dosimeter due to its high sensitivity to ionizing radiation because of the enhanced intensity of TSL (thermally stimulated luminescence) and OSL as against single doping (i.e. $KCl:Eu^{2+}$ and $KCl:Ce^{3+}$) [4]. It has been also reported that the occurrence of different states of dopants $Eu^{2+}$ due to the heat treatment influences the absorption and fluorescence optical properties for $KCl\text{-}KBr\text{-}RbBr:Eu^{2+}$ crystals [5]. Furthermore, it was recently found that $BaFBr_{0.85}I_{0.15}:Eu^{2+}$ demonstrates reusable and excellent dosimetry such as $KCl:Eu^{2+}$ [6].

By irradiating $KCl:Eu^{2+}$ crystal with the X-ray, electron–hole pairs are made in the crystal and chlorine vacancies store the electrons. This process generates *F*-centres (trapped electrons) and $F_z$-centres (*F*-centres near $Eu^{2+}$-cation vacancy dipoles). The electrons are liberated from the *F*-centres under a photostimulation of 560 nm, which agrees with the absorption wavelength of *F*-centre in $KCl:Eu^{2+}$ [7], and recombine



IntechOpen

with the trapped holes. Then, characteristic $Eu^{2+}$ luminescence results from the energy released in the PSL process of which the intensity becomes large with the irradiation dose of X-ray [8, 9].

The free electrons are released from the $F_z$-centre at 370 K and *F*-centre at 450 K and combine with $Eu^{3+}$ ions. Then, the two peaks are created on TL glow curve, which show the 420 nm emission due to the $4f^6 5d \rightarrow 4f^7$ transition of $Eu^{2+}$ ions [10].

A number of studies on OSL and TL processes have been carried out for X-ray-irradiated KCl:$Eu^{2+}$ and have proved that $F_z$- and *F*-centres are directly involved as the trapping sites for the electrons in their processes. But, to my knowledge, the details seem not to be clear (e.g. [11, 12]). In this chapter, a further insight into this situation is described with the help of optical absorption and TL glow measurements for the KCl:$Eu^{2+}$ single crystals after X-ray irradiation. KCl single crystals doped with $Eu^{2+}$ (0.02, 0.04, 0.1 mol% in melt) were prepared by cleaving the ingot, which was grown by the vertical Bridgman method, to the size of $2 \times 3 \times 5$ $mm^3$. The crystals were thereafter heat-treated in order to remove internal strains as much as possible and disperse the dopants in the crystal. The X-ray irradiation was conducted with W-target operated at 40 kV and 30 mA by Shimadzu XD-610. The absorption spectra were measured with a Hitachi U-3500 spectrophotometer. TL was observed using a photomultiplier (Hamamatsu R928) with a sensitive region from 185 to 900 nm. These were conducted at 300 to 540 K, and the irradiation and measurements were performed in the dark to avoid the unwanted effect of environmental light.

In the Section 4 of this chapter, absorption spectrum and TL glow curve are also reported for the crystals under the influence of compressive deformation after the irradiation. The crystals were plastically deformed by compression along the <100> axis direction at the cross-head speed of 100 μm $min^{-1}$ using an Instron 4465 testing machine at room temperature.

## 2. Absorption spectrum and TL glow curve for X-ray-irradiated crystals

Variations of absorption spectra and TL glow curves with X-ray irradiation times are shown in **Figure 1a** and **b** for the irradiated KCl:$Eu^{2+}$ (0.04 mol%) single crystals, which were irradiated by the X-ray up to the time of 120 min at room temperature.

Absorption coefficient is obtained from $(\log I_0 / I)/(d \log e)$, where $\log I_0 / I$ is the absorbance, and *d* is the width of specimen. The unit of absorption coefficient is $cm^{-1}$ according to the above description. This is theoretically based on the assumption that defects in solid (absorption source) are equally distributed in the solid. KCl:$Eu^{2+}$ single crystal before and after the X-ray irradiation is shown in **Figure 2a** and **b**.

In this study, the defects as absorption source were made by the X-ray irradiation at room temperature. This method cannot make the uniform distribution of defects in the specimen exposed to X-ray on the wide surfaces. The concentration distribution of *F*-centres (trapped electrons) was observed also for the X-ray-irradiated KBr single crystal in the article [13]. **Figure 3** and **Table 1** shows the variation of *F*-centre concentration with the distance from the surface of X-ray-irradiated KBr single crystal. As illustrated in **Figure 3**, the sample was exposed to X-ray for 3 h on each of the pair wide surfaces at room temperature and was cleaved in four thin crystal plates (a) ~ (d) at regular intervals (a thickness of about 1 mm). The *F*-centre concentration is smaller as the position of the specimen becomes centrally. Although the unit of absorption coefficient is supposed to be $cm^{-1}$ for an unirradiated specimen, the unit of it (i.e. the vertical scale of **Figure 1**) for the X-ray-irradiated specimen is accordingly represented by a.u. (arbitrary unit).

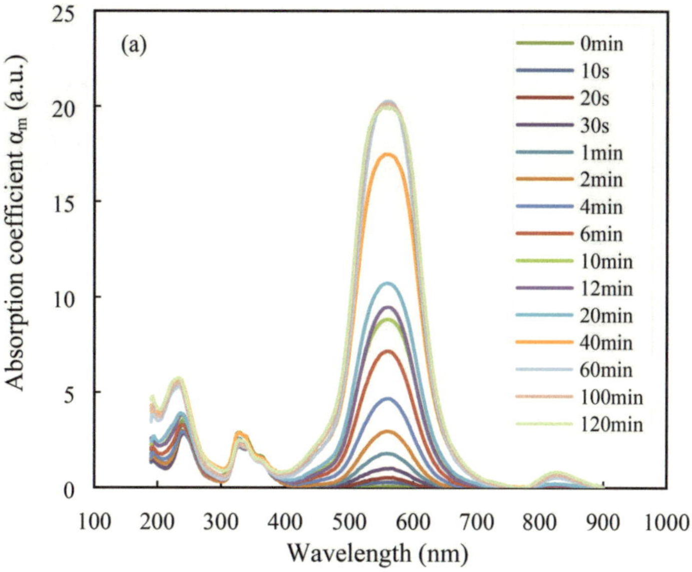


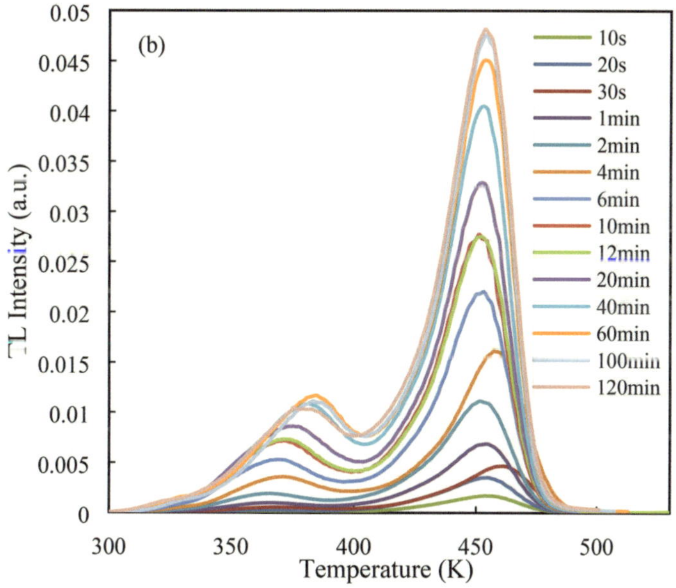


**Figure 1.**
*Variations of (a) absorption spectrum and (b) TL glow curve with the irradiation time for X-ray-irradiated KCl:Eu$^{2+}$ (0.04 mol%) single crystals.*

In **Figure 1a**, the absorption peaks at the wavelength of 243 and 343 nm are due to Eu$^{2+}$, and the peak at 560 nm is due to *F*-centre [7, 10]. A small absorption peak at 825 nm is ascribed to the *M*-centre, which is a complex centre comprising two neighboring Cl vacancies and a trapped electron [7]. **Figure 1b** shows TL glow curve of X-ray irradiated KCl:Eu$^{2+}$ (0.04 mol%), which was obtained by the linear heating at 24 K min$^{-1}$. The TL glow peak at the temperature of 370 K is related to $F_z$-centre (*F*-centre perturbed by neighboring a Eu$^{2+}$ ion-cation vacancy dipole), while that

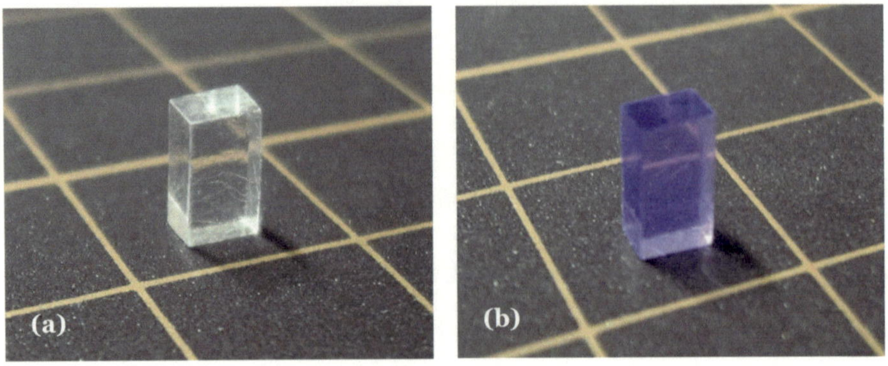


**Figure 2.**
*$KCl:Eu^{2+}$ single crystal (a) before and (b) after the X-ray irradiation.*

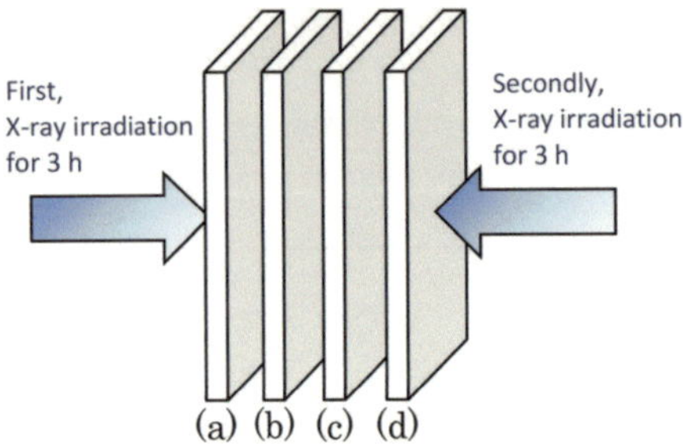


**Figure 3.**
*Four thin crystal plates (a)–(d) cleaved from X-ray-irradiated KBr single crystal.*

| Crystal plate | (a) | (b) | (c) | (d) |
|---|---|---|---|---|
| Concentration of *F*-centres ( $\times 10^{16} cm^{-3}$ ) | 25 | 3.0 | 3.3 | 28 |
| *Reproduced form Ref. [13].* | | | | |

**Table 1.**
*F-centres concentration in each crystal plate (a)–(d) cleaved from the X-ray-irradiated KBr single crystals.*

at 450 K seems to be due to *F*-centre [14, 15]. The former low-temperature peak and the later high one are termed LTP and HTP in this chapter, respectively. The height of absorption peak due to *F*-centre at 560 nm becomes larger as the irradiation time is longer in **Figure 1a**. This is caused by the increase in *F*-centre concentration with X-ray irradiation dose. The production of *F*-centre leads to the decrease of the concentration of the $Eu^{2+}$ ions in the crystal (because $Eu^{2+} \rightarrow Eu^{3+} + e^{-}$). But the peak height at the wavelength of 343 nm due to the $Eu^{2+}$ ion seems mostly unchanged taking account of the baseline, as shown in **Figure 1a**. This is because the *F*-centre concentration is much higher as against $Eu^{2+}$ ion in the crystals. As an instance, the concentration ($N$) of $Eu^{2+}$ is 105.6 ppm for 40 min X-ray-irradiated $KCl:Eu^{2+}$ (0.04 mol% in melt) single crystal according to Hernandez et al. [16] from the absorption peak $\alpha_m$ = 3.0 at 243 nm in **Figure 1a**, namely.

$$N = 35.2\,\alpha_m \tag{1}$$

While, the concentration ($N_0$) of *F*-centre is 2.57 ppm for the above-mentioned specimen using Smakula's formula given by [17].

$$N_0 = \frac{1}{2.06} 10^{16} \frac{\alpha_m}{f} \left( \frac{r}{10^8} \right)^3 \tag{2}$$

where $f$ = 0.8 is the oscillator strength, absorption peak $\alpha_m$ = 17.479 at 560 nm in **Figure 1a**, and $r$ = 3.1464 Å [18] is the interionic distance of KCl at room temperature. While, the absorption peak at about 825 nm due to the *M*-centre is insensitive to the increment of X-ray irradiation dose.

In **Figure 1b**, the heights of both LTP (due to $F_z$ centre) and HTP (due to *F*-centre) increase with the irradiation time. And also, LTP shifts to slightly higher temperature with it. This may suggest that the excitonic electrons are trapped into deeper states when the irradiation dose increases. HTP lies a little towards high-temperature side below 6 min of the time.

## 3. F-bleach and thermal-bleach effects on F-centre peak, LTP, and HTP

The absorption spectra and the TL intensity of X-ray irradiated KCl:$Eu^{2+}$ were measured after *F*-light (560 nm wavelength) bleach by using a Hitachi U-3010 fluorescence spectrometer after the irradiation. The *F*-light source was a xenon lamp. **Figure 4a** and **b** represents the absorption coefficient $\alpha_m$ at 560 nm wavelength, LTP, and HTP of dependence on the *F*-light exposure time for 8 min X-ray-irradiated KCl:$Eu^{2+}$ (0.1 mol%) single crystals. The *F*-light bleach was carried out with an optical bandpass filter of 5 nm bandwidth. The *F*-bleach was done for the purpose of removing the *F*-centre peak within the bandwidth from the TL signal. As a result, the $F_z$-centre peak is enhanced. This may lead to resolve the $F_z$-centre.

In **Figure 4a** and **b**, the values of $\alpha_m$ at 560 nm and HTP decrease with increasing *F*-light exposure time. By the *F*-light bleaching, the *F*-centre peak is removed within 5 nm bandwidth from the TL signal. But LTP is almost constant independently of the exposure time as shown in **Figure 4b**.

In addition, the X-ray irradiated crystals were similarly done using *F*-light at the excitation bandpass of 20 nm. The result of *F*-bleach is shown in **Figure 4c** and **d** for the irradiated crystals for 6 min. Not only the values of $\alpha_m$ at 560 nm and HTP but also LTP decreases with the *F*-light exposure time. This means the hidden peak due to the $F_z$-centre is either on the lower or on the higher side of 560 nm in the absorption spectrum. In **Figure 4b** and **d**, LTP initially increases in intensity below the exposure time 1 min. This is because $F_z$-centres are formed by the dynamical rearrangement of *F*-centres to $Eu^{2+}$ as described in the papers [12, 15]. That is to say, *F*-centres transform partially to $F_z$-centres during the *F*-bleach.

The thermal bleach was further carried out for 10 min X-ray-irradiated KCl:$Eu^{2+}$ (0.02 mol%) single crystals. The absorption spectra and the TL intensity of the irradiated KCl:$Eu^{2+}$ were also measured by using a Hitachi U-3010 fluorescence spectrometer after the thermal bleach. The thermal bleach was carried out by heating at the rate of 24 K $min^{-1}$ up to 523 K after the irradiation. The bleach temperature dependence of *F*-centre peak in absorption spectrum, LTP, and HTP is shown in **Figure 5**.

The LTP height (blue solid line) decreases with increasing the bleach temperature above 360 K and gradually approaches to zero above 400 K. *F*-centre peak (red solid one) also begins to become small with it at the same thermal-bleach temperature

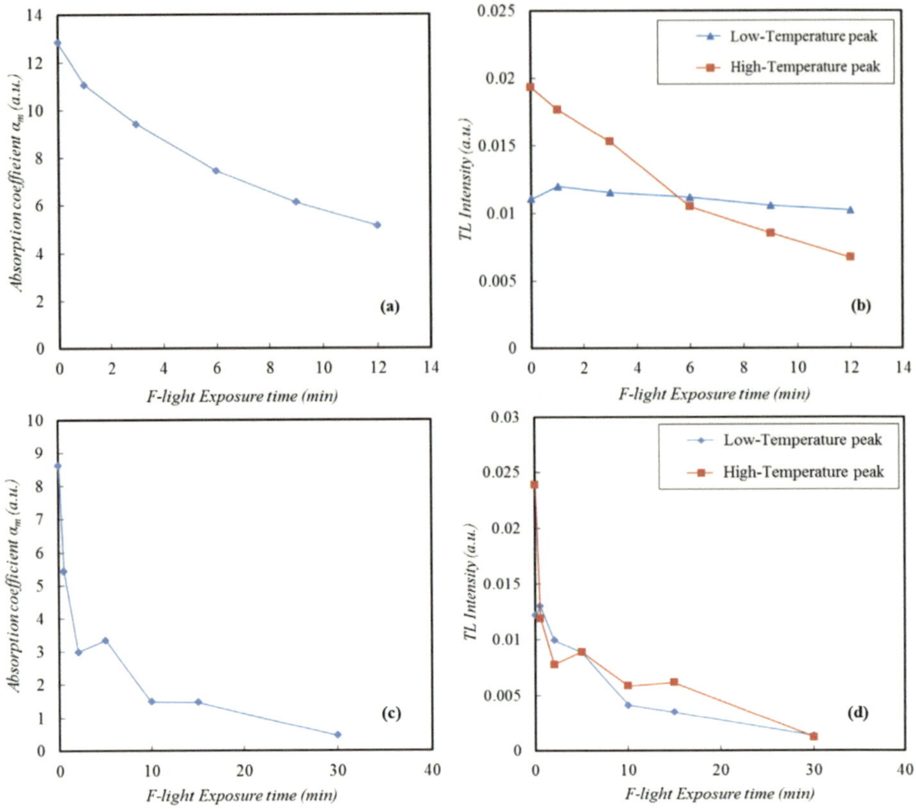


**Figure 4.**
*Relations between F-light exposure time versus the absorption coefficient and each peak (LTP, HTP) of TL glow curve for X-ray- irradiated KCl:Eu$^{2+}$ (0.1 mol%) single crystals. The X-ray irradiation time and the excitation bandpass: (a) (b) 8 min and 5 nm; (c) (d) 6 min and 20 nm [19].*

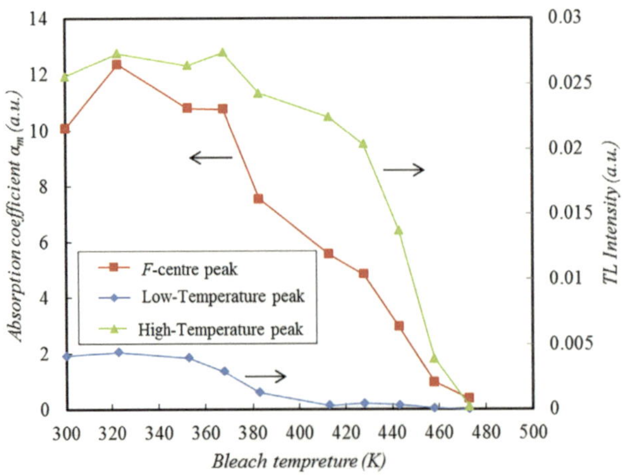


**Figure 5.**
*Relations between thermal-bleach temperature versus F-centre peak in absorption spectrum and each peak (LTP, HTP) of TL glow curve for X-ray-irradiated (irradiation time: 10 min) KCl:Eu$^{2+}$ (0.02 mol%) single crystals [19].*

(360 K). This suggests that $F_z$- and $F$-centres peaks are at nearly equal wavelength, as the above-mentioned results of $F$-light bleach. $F$-centre peak continually decreases with the bleach temperature at 400–473 K in accordance with HTP (green solid line) of dependence on the bleach temperature. On the basis of the results of $F$-light and thermal bleaches for the crystals, it is considered that electrons trapped at shallow traps (i.e. $F_z$-centres) begin to be set free at about 370 K and end detrapping from the traps around 400 K. The electrons trapped at deep traps (i.e. $F$-centres) begin to detrap continuously at 400–473 K. $F_z$-centre peak height is so low that it cannot be observed here. The two absorption peaks due to $F$- and $F_z$-centres are considered to be overlapped around the peak at wavelength 560 nm in the absorption spectrum for the crystals.

## 4. Optical absorption and TL glow peaks under the influence of compressive deformation

**Figure 6a** and **b** shows the absorption spectra and TL glow curves for 10 min X-ray-irradiated KCl:$Eu^{2+}$ (0.02 mol%) single crystal before (blue solid line) and after (red solid one) the compression (applied stress: 20.8 MPa), where luminescence intensity and load signals were recorded on a computer and then the data were processed.

The peak at the wavelength of 560 nm due to $F$-centre [7, 10] appears on the absorption spectra irrespectively of the compression. As described in Section 3, it is considered that the peak due to $F_z$-centre is approximately within 20 nm of the wavelength 560 nm in the absorption spectrum of X-ray-irradiated KCl:$Eu^{2+}$ single crystal. The height of the peak around 560 nm becomes larger under the influence of compression, as can be seen in **Figure 6a**. This is assumed to be the phenomenon that many $F$ and $F_L$-centres in the crystal are expected to be formed increasing vacancy during the process of compressive deformation.

On the other hand, the TL glow curve has two peaks (i.e. LTP and HTP) at the temperatures of 370 and 450 K for the X-ray-irradiated crystal before the compression. After the compressive deformation for the irradiated crystal, a new peak appears around 330 K on the TL glow curve in **Figure 6b**. This peak is termed 330 K peak in this chapter.

As for pure KCl single crystal after the deformation, there are no additional bands characteristic of local radiation defects on absorption spectrum [21].

**Figures** 7 and **8** show the influence of 330 K thermal bleaching on the absorption spectra and TL glow curves for the compressed and subsequent X-ray-irradiated KCl:$Eu^{2+}$ (0.02 mol%) crystal. Blue and red solid lines in these figures are related to the crystals before and after the thermal-bleaching treatment, respectively. The 330 K thermal-bleach time was 35 min and X-ray irradiation time 10 min. In addition, the absorption spectrum for the irradiated crystal before the compression is represented by a green solid line in **Figure** 7.

By the 330 K thermal bleaching with 5 nm bandpass filter, the peak at the wavelength of 560 nm due to $F$-centre decays until near height of the irradiated crystal before the compression in the absorption spectra, as shown in **Figure** 7. Then, the 330 K peak disappears on the TL glow curve in **Figure 8**. TL intensities of LTP and HTP are almost constant as expected. On the basis of these results on 330 K thermal bleach, it is considered that the increase in $F$-centre peak after the compression in **Figure** 7 is attributable to the production of the color centres (330 K color centre) caused the 330 K peak for the compressed crystal. The absorption wavelength at

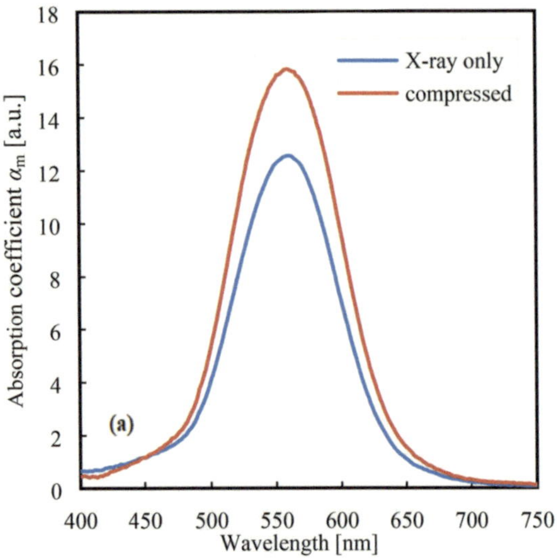


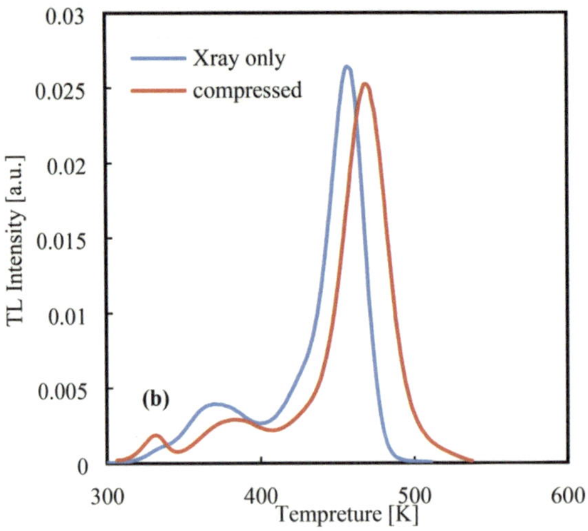


**Figure 6.**
*(a) Absorption spectra and (b) TL glow curves for X-ray-irradiated KCl:Eu$^{2+}$ (0.02 mol%) single crystals after the following treatments: (—) X-ray irradiation; (—) X-ray irradiation → compression (reproduced from Ref. [20] with permission from the publisher).*

$F_Z$-centre peak is supposed to be approximately within 20 nm of that (560 nm) at *F*-centre peak (see Section 3). And further, the 330 K color centre may be also overlapped near 560 nm in the absorption spectrum of compressed crystal (blue solid line) from **Figure** 7. Here, it was investigated whether the height of 330 K peak is influenced by bleaching with *F*-light (bandpass 5 nm) for 10 min X-ray-irradiated KCl:Eu$^{2+}$ (0.02 mol%) single crystals after the compression (applied stress: 20.8 MPa). **Figure 9** shows the *F*-bleaching results, where red and blue solid lines represent the 330 K peak of TL glow curve and *F*-peak in the absorption spectrum, respectively.

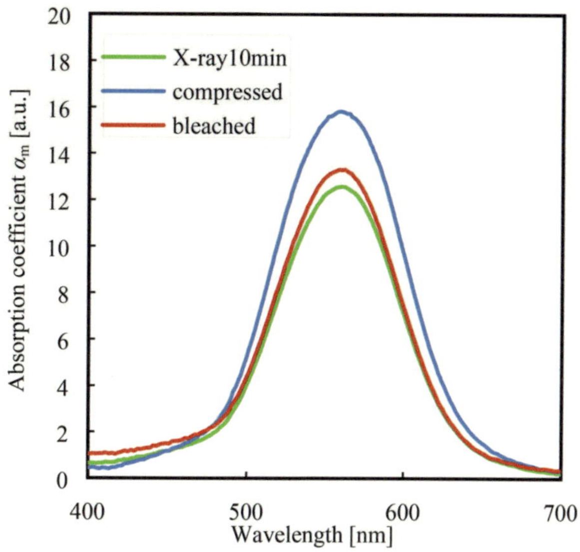


**Figure 7.**
*Absorption spectra for KCl:$Eu^{2+}$ (0.02 mol%) single crystals after the following treatments: (—) X-ray irradiation; (—) X-ray irradiation→compression→X-ray irradiation; (—) X-ray irradiation→compression→X-ray irradiation→thermal-bleach at 330 K for 35 min. X-ray-irradiation time was 10 min (reproduced from Ref. [20] with permission from the publisher).*

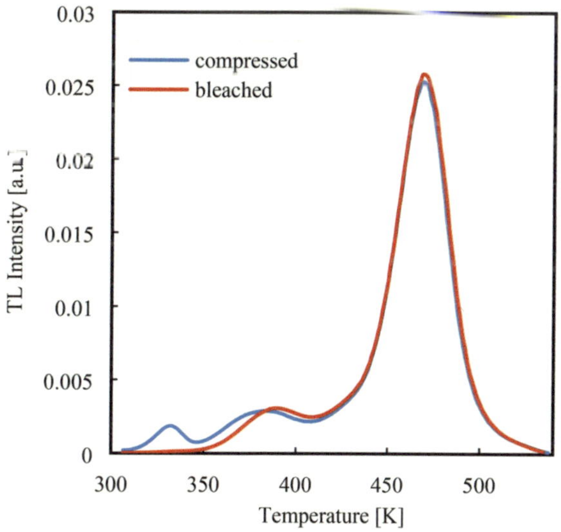


**Figure 8.**
*TL glow curves for KCl:$Eu^{2+}$ (0.02 mol%) single crystals after the following treatments: (—) X-ray irradiation→compression→X-ray irradiation; (—) X-ray irradiation→compression→X-ray irradiation→thermal-bleach at 330 K for 35 min. X-ray-irradiation time was 10 min. (reproduced from Ref. [20] with permission from the publisher).*

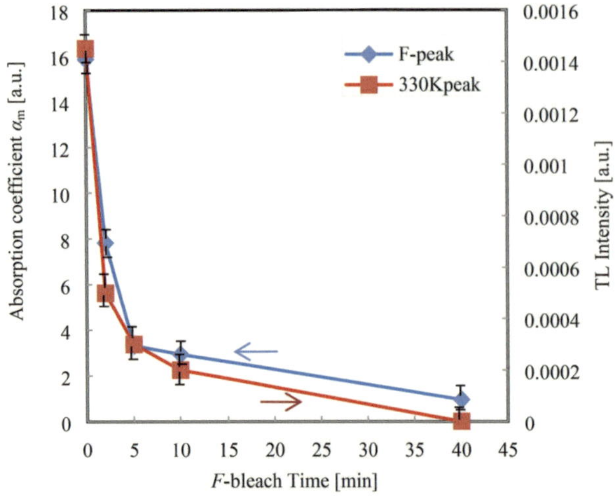


**Figure 9.**
*Relations between F-bleach time (excitation bandpass: 5 nm) versus (—) 330 K peak of TL glow curve and (—) F-centre peak in absorption spectrum for KCl:Eu$^{2+}$ (0.02 mol%) single crystals previously compressed and subsequently X-ray-irradiated (irradiation time: 10 min). (reproduced from Ref. [20] with permission from the publisher).*

The TL intensity of 330 K peak rapidly decreases with increasing *F*-light exposure time until about 5 min and gradually approaches to zero in accordance with the variation of *F*-peak. Therefore, it is considered that the absorption wavelength due to the 330 K color centre is near 560 nm in the absorption spectrum.

## 5. Conclusion

The X-ray irradiation induces $F_Z$- and *F*-centres in KCl:Eu$^{2+}$ crystal, of which the wavelengths are almost equal with each other in the absorption spectrum. The absorption peak due to the $F_Z$-centre is either on lower or higher side within 20 nm of the wavelength 560 nm in the absorption spectrum.

The changes in optical absorption and TL glow curves induced by compressive deformation of KCl:Eu$^{2+}$ were presented here. That is, the new peak (i.e. 330 K peak) appears around 330 K on TL glow curve. The analyses of *F*-light bleach and 330 K thermal bleach reveal that the absorption wavelength due to the 330 K peak is near the *F*-centre peak in the absorption spectrum for the crystal.

## Acknowledgements

The author is pleased to give thanks to Dr. T. Ohgaku for helpful and constructive discussions on the article and to Y. Igari for his experimental assistance.

## Conflict of interest

The author declares no conflict of interest.

## Author details

Yohichi Kohzuki
Department of Mechanical Engineering, Saitama Institute of Technology, Fukaya, Japan

*Address all correspondence to: kohzuki@sit.ac.jp

## References


[1] Camacho AQ, Muñoz GH, Rubio JO, Garcia JM, Murrieta HS, Hernandez JA. Dosimetric properties of KCI:Eu. Journal of Materials Science Letters. 1988;7:437-440. DOI: 10.1007/BF01730681

[2] Driewer JP, Chen H, Osvet A, Low DA, Li H. Radiation hardness of the storage phosphor europium doped potassium chloride for radiation therapy dosimetry. Medical Physics. 2011;**38**:4681-4688. DOI: 10.1118/1.3611043

[3] Hansel RA, Xiao Z, Zhang L, Li HH. X-ray storage performance of KCl:$Eu^{2+}$ with high cumulated dose. Nuclear Instruments and Methods in Physics Research Section B: Beam Interactions with Materials and Atoms. 2014;**326**:178-181. DOI: 10.1016/j.nimb.2013.08.067

[4] Krishnakumar DN, Rajesh NP. Growth and optical characterization of europium and cerium doped KCl single crystals by Czochralski method for dosimetric applications. Journal of Electronic Materials. 2019;**48**(3):1629-1633. DOI: 10.1007/s11664-018-06863-3

[5] Martínez-González JU, Cordero-Borboa AE, Hernández-Juárez E. Optical properties of the $Eu^{2+}$-ion immersed as a doping impurity into a KCl:KBr:RbBr solid solution: Influence of the different aggregation-precipitation states of the doping ions. Solid State Communications. 2019;**300**:113671. DOI: 10.1016/j.ssc.2019.113671

[6] Setianegara J, Mazur TR, Yang D, Li HH. Dual-storage phosphor proton therapy dosimetry: Simultaneous quantification of dose and linear energy transfer. Medical Physics. 2021;**48**(4):1941-1955. DOI: 10.1002/mp.14748

[7] Kao KJ, Perlman MM. X-ray effects on cation impurity-vacancy pairs in KCl:$Eu^{2+}$. Physical Review B. 1979;**19**:1196-1202. DOI: 10.1103/PhysRevB.19.1196

[8] Nanto H, Murayama K, Usuda T, Taniguchi S, Takeuchi N. Optically stimulated luminescence in KCl:Eu single crystals. Radiation Protection Dosimetry. 1993;**47**:281-284. DOI: 10.1093/oxfordjournals.rpd.a081751

[9] Nanto H, Ikeda M, Nishishita J, Kadota M, Nasu S, Douguchi Y, et al. Eu concentration dependence of photostimulated luminescence in x-ray or uv-ray irradiated KCl:Eu phosphor ceramics. Nuclear Instruments and Methods in Physics Research Section B: Beam Interactions with Materials and Atoms. 1996;**116**(1-4):542-544. DOI: 10.1016/0168-583X(96)00104-8

[10] Rubio JO, Flores MC, Murrieta HS, Hernandez JA, Jaque F, Lopez FJ. Influence of concentration and aggregation-precipitation state of divalent europium in the room-temperature coloring of KCl. Physical Review B. 1982;**26**:2199-2207. DOI: 10.1103/PhysRevB.26.2199

[11] Pedroza-Montero M, Castañeda B, Meléndrez R, Piters TM, Barboza-Flores M. Thermoluminescence, optical stimulated luminescence and defect creation in europium doped KCl and KBr crystals. Physical Status Solidi (b). 2000;**220**:671-676. DOI: 10.1002/1521-3951(200007)220:1<671::AID-PSSB671>3.0.CO;2-L

[12] Chernov V, AoR M, Piters TM, Barboza-Flores M. Thermally and optically stimulated luminescence correlated processes in x-ray irradiated KCl:$Eu^{2+}$. Radiation Measurements. 2001;**33**:797-800. DOI: 10.1016/S1350-4487(01)00110-X

[13] Kohzuki Y. Dislocation-point defects-induced by X-irradiation interaction in alkali halide crystals. In: René M, editor. Recent Advances in Mineralogy. London, UK, London: IntechOpen; 2023. DOI: org/10.5772/intechopen.107567

[14] Aceves R, Pérez Salas R, Barboza-Flores M. The role of F centres in the thermoluminescence of low-energy UV- and x-irradiated KCl:$Eu^{2+}$. Journal of Physics: Condensed Matter. 1994;**6**:10397-10405. DOI: 10.1088/0953-8984/6/47/022

[15] Pedroza-Montero M, Castañteda B, Meléndrez R, ChernovV, Barboza-FloresM. Comparative investigations of TL and OSL in KCI:$Eu^{2+}$ crystals irradiated with UV and X-rays. Radiation Effects and Defects in Solids. 2001;**154**:319-324. DOI: 10.1080/10420150108214071

[16] Hernandez AJ, Cory WK, Rubio OJ. A non-destructive method for determining the $Eu^{2+}$ concentration in the alkali chlorides. Japanese Journal of Applied Physics. 1979;**18**:533-538. DOI: 10.1143/JJAP.18.533

[17] Dexter DL. Theory of the optical properties of imperfections in nonmetals. Solid State Physics. 1958;**6**:353-411. DOI: https://doi.org/10.1016/S0081-1947(08)60730-4

[18] Sirdeshmukh DB, Sirdeshmukh L, Subhadra KG. Alkali Halides. Berlin Heidelberg: Springer-Verlag; 2001. p. 7. DOI: 10.1007/978-3-662-04341-7

[19] Kohzuki Y, Ohgaku T. Study on luminescence of KCl:$Eu^{2+}$ crystals after X-ray irradiation at room temperature. Crystals. 2019;**9**(7):331. DOI: 10.3390/cryst9070331

[20] Kohzuki Y, Ohgaku T. Deformation luminescence of X-ray-irradiated KCl:$Eu^{2+}$ single crystals by compression. Journal of Luminescence. 2023;**253**:119469. DOI: 10.1016/j.jlumin.2022.119469

[21] Shunkeyev K, Myasnikova L, Maratova A, Bizhanova K. Mechanisms of radiation defect formation in the KI crystal in the deformation field. In: 7th International Congress on Energy Fluxes and Radiation Effects (EFRE). IEEE; 2020. pp. 1016-1020. DOI: 10.1109/EFRE47760.2020.9242132

Chapter 2

# Fundamental Concerns of Optical Fluorescence Intensity Ratio-Based Thermometry

*Helena Cristina Vasconcelos*

## Abstract

This chapter provides a comprehensive exploration of optical fluorescence intensity ratio (FIR) temperature sensing, blending theoretical underpinnings with practical applications. It underscores the intrinsic sensitivity and non-invasiveness of FIR technology, spanning diverse scientific disciplines where its utility is paramount. Central to the discussion are the intricate energy transfer mechanisms within fluorescence emissions from temperature-sensitive materials, revealing their nuanced responses to thermal changes. Fundamental to FIR thermometry are the lanthanide ($Ln^{3+}$) ions, which play pivotal roles due to their unique electronic configurations. These elements exhibit temperature-dependent variations in fluorescence properties, including intensity and lifetime, crucial for accurate temperature determination. Specifically, the chapter delves into the utilization of erbium ($Er^{3+}$) and holmium ($Ho^{3+}$) ions in the context of FIR thermometry, highlighting their distinct contributions to enhancing temperature sensitivity. The $Er^{3+}/Ho^{3+}$ co-doped nano-garnet emerges as a promising material in this field, effectively bridging theoretical frameworks with practical implementations. The narrative is enriched by the incorporation of the Boltzmann distribution equation, which provides a robust theoretical foundation for understanding temperature-dependent fluorescence phenomena exhibited by $Ln^{3+}$ ions. This chapter serves as a valuable resource, offering a concise understanding on the forefront of optical FIR-based thermometry for researchers and professionals alike.

**Keywords:** lanthanide ions ($Ln^{+3}$), temperature sensitivity, luminescence, optical FIR-based thermometry, fluorescence intensity ratio (FIR)

## 1. Introduction

The foundations of temperature measurement date back to the sixteenth and seventeenth centuries, a crucial period in the history of scientific instrumentation. During this era, Galileo made significant contributions by introducing the thermoscope [1]. This early device served as a forerunner to modern temperature-measuring instruments. Although these early instruments provided reasonable precision, they suffered from a lack of reproducibility [2]. Despite these limitations, the thermoscope paved the way for future advancements in thermometry. The term "thermometer" emerged in 1624, marking a major milestone in the development of temperature-measuring instruments [3].

Building on this legacy, temperature—a fundamental property of matter—can now be measured using various types of thermometers. These modern instruments, which are macroscopic systems that respond to changes in temperature, utilize different principles, such as thermal expansion, gas pressure, resistance, and infrared radiation, to provide accurate and reliable temperature measurements across a variety of applications. Indeed, one common approach relies on the thermal expansion of a liquid, leveraging the principle that liquids expand when heated and contract when cooled. Another method involves using the expansion of a gas, which can measure temperature accurately in one of the two keyways: either maintaining a constant pressure or maintaining a constant volume. These gas-based thermometers operate according to the ideal gas law, which offers a robust foundation for temperature measurement.

In addition to these methods, other thermometers utilize the temperature dependence of electrical resistance. As temperature changes, the resistance of certain materials also changes, providing a means to measure temperature. Furthermore, thermometers can also use black body radiation to measure temperatures accurately in specific scenarios. This approach involves analyzing the spectrum of radiation emitted by a body to determine its temperature. A physical measurement is achieved by comparing the physical magnitude of a body with a standard. To accurately measure something, it is important to understand the basic principle behind it, which is the physical phenomenon the measurement relies on. In luminescence thermometry, this principle relies on the change in any parameter of the luminescence emitted by a material when it is exposed to different temperatures.

Lord Kelvin, also known as William Thomson, recognized the importance of aligning temperature measurement with the kinetic theory of gases. In 1848, he proposed the Kelvin scale, establishing absolute zero as its starting point [4]. This scale is widely used, particularly in thermodynamics and physics, providing precise temperature measurements essential for accurate calculations. According to the International System of Units (SI), the Kelvin scale is the official temperature scale. Since 2019, the definition of the Kelvin has been based on the Boltzmann constant rather than the water triple point.

Boltzmann's constant, a pivotal concept, establishes a crucial link between temperature and energy. It is integrated into the ideal gas law and serves as a conversion factor between Kelvin and Joules. This constant underscores the inherent relationship between temperature and energy, revealing their intimate connection. By understanding these concepts, one gains insights into the intricate nature of temperature measurement and its profound implications in the realm of thermodynamics. Nevertheless, the concept of temperature is closely tied to our subjective sensory perception of an object's heat or coldness. In simple terms, temperature is the measure of how hot or cold a body is, expressed in various temperature scales, such as Fahrenheit, Celsius, or Kelvin.

When considering a substance in a classical sense, with its constituent molecules continuously moving in a random and chaotic manner due to thermal energy, temperature is often defined as a measure of the translational kinetic energy stemming from this disorderly motion [5]. Higher temperatures correlate with more significant thermal movement. In simple terms, in accordance with the classical statistical physics, the average kinetic energy per particle is related to thermal motion. From a physics perspective, temperature represents the average kinetic energy of the atoms or molecules in a system. Essentially, temperature is the external indicator of the thermal energy present within a system. Absolute zero (0 K) is the lowest possible temperature, where matter contains no thermal energy. By measuring the temperature difference between two substances, we can identify the direction of heat flow, which always travels from a region of higher temperature to one of lower temperature. The equipartition theorem delineates the intricate relationship between temperature and the contributions of individual degrees of freedom within a system. Therefore, the total average kinetic energy of a molecule in an ideal gas would be

$$\frac{1}{2}kT + \frac{1}{2}kT + \frac{1}{2}kT = \frac{3}{2}kT \tag{1}$$

Each degree of freedom contributes $\frac{1}{2}kT$ to the total energy, with $k$ representing the Boltzmann constant and $T$ is the temperature [5]. This connection reveals the profound influence of temperature on the microscopic behaviors exhibited by particles in a substance. When two systems share the same temperature, they exist in a state of thermal equilibrium, signifying an absence of net thermal energy transfer between them. Conversely, if one system is at a higher temperature than another, energy will flow spontaneously from the hotter system to the colder one when they come into thermal contact [5]. Broadening our perspective to the macroscopic scale, the second law of thermodynamics emerges as a guiding principle. This law dictates that the total entropy of an isolated system tends to increase over time during spontaneous processes. Entropy, often described as a measure of disorder or randomness within a system, becomes intimately connected to temperature through the relation ship expressed by the equation:

$$\Delta S = \frac{\Delta Q}{T} \tag{2}$$

where $\Delta S$ represents the change in entropy, $\Delta Q$ signifies heat transfer, and $T$ stands for the absolute temperature. Through the blending of these concepts, the temperature serves as the bridge connecting the statistical behaviors of particles to the overarching principles governing energy exchange in physical systems [5].

Recognized as a fundamental variable in science, temperature plays a pivotal role in various industrial and scientific processes [1, 6]. It is essential for controlling chemical reactions, manufacturing processes, and energy production, among other applications. Accurate temperature measurement and control are vital for ensuring the quality and safety of products and for optimizing efficiency in systems. The significance of temperature spans across multiple fields. In the medical field, temperature is a critical factor for diagnosing and monitoring patients. Body temperature is a key indicator of health, and fluctuations can signal the presence of infection, inflammation, or other medical conditions. Temperature also plays a role in preserving and transporting biological samples, vaccines, and other medical supplies. Additionally, it is essential in various medical procedures, such as thermotherapy and cryotherapy, where precise temperature control is necessary for effective treatment. Understanding and managing temperature in medical settings is crucial for patient care and overall health outcomes.

In scientific research, temperature plays a significant role in the study of various physical and chemical phenomena. It influences the behavior of materials at the atomic and molecular levels, as well as the properties of substances across different phases.

The evolution of temperature measurement, from Galileo's thermoscope to modern methodologies, traces a rich historical journey underscored by continual innovation. Overcoming the limitations of early thermoscopes has spurred the development of diverse thermometer technologies, culminating in the introduction of non-contact measurement techniques. This progression exemplifies the dynamic nature of temperature measurement, driven by technological advancements.

As technology has advanced, novel techniques have emerged to enhance the accuracy and precision of temperature readings. Among these modern methods is luminescence thermometry, leveraging the luminescent properties of materials to gauge temperature. This approach marks a significant departure from traditional methods, offering non-contact and high-resolution temperature measurement capabilities, thus ushering in a new era in temperature sensing technology. By observing the changes in luminescence intensity or wavelength in response to temperature variations, researchers and engineers can determine the temperature of a system with great accuracy. This method builds on the legacy of temperature measurement, linking the past to the present, by incorporating principles of light and energy that Galileo and his contemporaries explored. Luminescence thermometry is particularly useful in challenging environments such as high-temperature or remote locations, where traditional methods may be less effective.

## 1.1 Luminescence thermometry

Traditional contact methods encounter difficulties in precisely measuring a material's temperature, mainly because introduced thermometers may cause interference and disturb the accurate evaluation of the material's temperature [7]. Conventional temperature measurement tools, such as bimetallic and liquid/gas-based thermometers, pyrometers, and thermocouples, among others, lack the precision needed for high spatial resolution, especially at scales on the order of micrometers, which are typical in cellular systems.

Optical sensors have noticeable advantages compared to conventional methods [8]. In recent years, researchers have become increasingly interested in optical thermometry, a sophisticated method that examines the relationship between temperature and various optical parameters. This modern approach offers key advantages, such as high spatial resolution, fast response, and non-invasive measurement. These benefits make it particularly useful in challenging environments like submicron scales, high-voltage areas, and caustic conditions [9]. Consequently, optical thermometers have become essential tools for precise temperature measurements in demanding settings.

Luminescence thermometry is a non-contact technique that measures temperature by observing the luminescent (light-emitting) properties of materials. This method is particularly beneficial when traditional temperature measurement techniques are impractical or impossible [10]. Some elements have optical properties that make them suitable for spectroscopic studies. $Ln^{3+}$ ions, which display luminescence properties like fluorescence emissions, are commonly used as dopants to aid in examining host materials due to their high efficiency as emitters [11]. This has promoted the development of new concepts for sensors based on the luminescence properties of materials, leading to the creation of optical sensors [10]. Therefore, $Ln^{3+}$ has attracted significant interest as optical temperature sensors because their fluorescence intensity can change in response to variations in their absorption and emission properties with temperature. These variations in fluorescence properties are examined using different techniques and are described by parameters, such as spectral position, bandwidth, intensity, decay lifetime, and fluorescence intensity ratio (FIR) [12].

Luminescence thermometry involves the emission of light when materials are electronically excited by an external energy source, such as optical radiation (in the case of photoluminescence) [10, 11]. This process is common in materials, such as dyes, semiconductors, and phosphors. The characteristics of the emitted light, including spectrum shape, bandwidth, and spectral shift, are influenced by the local temperature of the material. The technique leverages the complex interaction between temperature and luminescence, enabling thermal sensing through careful spatial and spectral analysis of emitted light as a material undergoes temperature changes. It provides a comprehensive solution, featuring high thermal resolution ($<0.1°C$), significant relative thermal sensitivity ($>1\%/°C$), and precise temperature mapping with exceptional optical spatial resolution ($<10\ \mu m$) [13]. Among the optical parameters used in temperature sensing, the fluorescence intensity ratio (FIR) stands out as the most reliable technique.

The key to developing an effective optical temperature sensor lies in selecting the right optically active ion and host matrix. Incorporating $Ln^{3+}$ ions is crucial due to their photostability and spectroscopic advantages, such as distinct emission spectra, extended fluorescence lifetimes, and high quantum yields [11]. Known for their strong luminescence and intricate energy level structures, $Ln^{3+}$ ions often exhibit narrow energy gaps between levels. However, not all of them are suitable for calibrating optical responses. Studies by Zhou et al. [14, 15] and Soler-Carracedo et al. [16] demonstrate how these features collectively enhance the precision and sensitivity of optical temperature measurements.

Considerations include the need for a delicate balance in the energy gap between thermalized levels, ensuring it is large enough to prevent emission overlap yet also allowing for a minimum upper-level population within the desired temperature range. Additionally, radiative probabilities associated with thermalized levels should be sufficiently high to produce significant emission intensities, and the relative intensities of emission peaks from different energy levels can be directly related to temperature. Praseodymium ($Pr^{3+}$), neodymium ($Nd^{3+}$), samarium ($Sm^{3+}$), thulium ($Tm^{3+}$), europium ($Eu^{3+}$), holmium ($Ho^{3+}$), and erbium ($Er^{3+}$) ions have energy level pairs suitable for optical temperature sensors. Berthou et al. [17] examined $Er^{3+}$ thermalized levels in fluoroindate fibers using the FIR technique across a large range of temperatures. Subsequent studies explored $Ln^{3+}$ ions and different host materials, such as $Er^{3+}$ or $Er^{3+}/Yb^{3+}$-doped various types of glasses, such as tellurite, fluorotellurite, oxyfluoride, fluoroindate, chalcogenide, and fluorophosphate glasses, for optical luminescent temperature sensing [18].

Garnets possess remarkable physicochemical properties, including hardness, high optical transparency, and strong mechanical and chemical stability. Their structure has been used as a host matrix for $Ln^{3+}$ ions [16]. Optically, the combination of the luminescence properties of $Ln^{3+}$ and the crystal stability of garnets makes them ideal for the development of new sensors in the context of nanothermometry.

Radiative decay processes allow for the precise measurement of these intensities. For example, in the $Nd^{3+}$-doped $Y_3Ga_5O_{12}$ nano-garnets, changes in the relative intensities of emission peaks between Stark levels of the $^4F_{3/2} \rightarrow {}^4I_{9/2}$ transition are used to measure temperature accurately [19]. Radiative decay is preferred over non-radiative decay because non-radiative processes (such as phonon emission or energy transfer to surrounding molecules) result in energy loss as heat rather than useful light emission. High radiative decay rates minimize these losses, improving the performance of the material in its intended application.

### 1.2 Fluorescence intensity ratio (FIR)

The way different luminescent properties—such as lifetime, intensity, and spectral characteristics—respond to temperature changes, as depicted in **Figure 1**, enables researchers to measure temperature. They do this by analyzing the emitted light from a material when it is exposed to an external excitation source.

One key aspect of luminescent thermometry is the change in luminescence lifetime [11]. Typically, as the temperature of a material increases, its luminescence lifetime decreases, resulting in a faster emission of light. Another important characteristic is the spectral changes that occur with temperature shifts. For instance, it can be observed as a decrease in luminescence intensity, a shift in peak wavelengths, or changes in the emission band as the temperature varies. However, one of the most effective and commonly used techniques in luminescent thermometry is based on the luminescence intensity ratio between two emission peaks [10, 16]. This method is particularly sensitive and widely applied in practice. By observing the changes in the

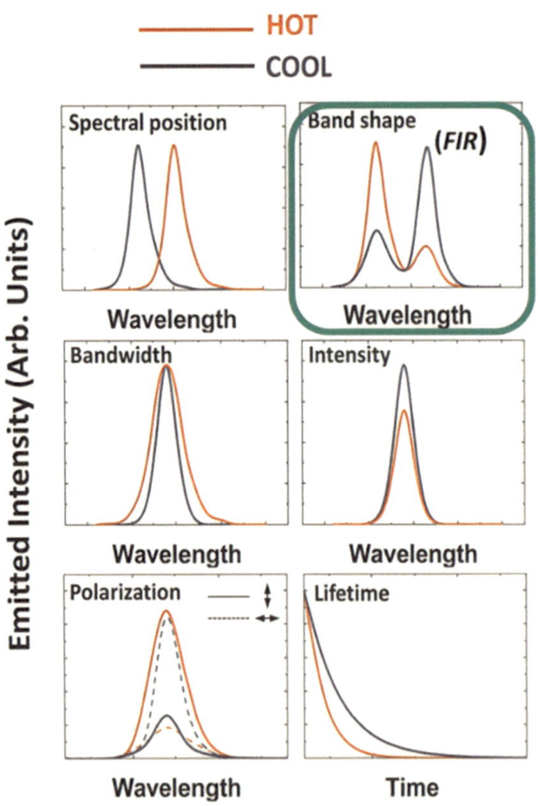


**Figure 1.**
*Diagram illustrating the potential impacts of an increase in temperature on luminescence. Adapted from Ref. [12].*

intensity ratio between two distinct peaks in the emission spectrum of luminescent ions, researchers can accurately determine temperature variations. When these ions are excited, they emit light from two energy levels separated by an energy difference (ΔE), designated as thermal coupling levels (TCLs).

The FIR technique offers a significant advantage by using the intensities from two closely spaced energy levels, known as thermally coupled energy levels (TCLs), to monitor temperature. The intensity ratio changes with temperature are independent of the source power since any variations in excitation power affect both levels equally. This enhances the measurement sensitivity and sensor stability. TCLs are in a thermodynamically quasi-equilibrium state, offering several benefits over non-coupled levels [20, 21]. First, the population of individual TCLs is directly proportional to the total population. Second, the theory of relative changes in fluorescence intensity from the two TCLs is well understood, making their behavior easier to predict. For FIR techniques to be effective, the energy gap between TCLs should be within the range of about 200–2000 $cm^{-1}$, which is satisfied by many rare-earth ions, such as $Tm^{3+}$, $Pr^{3+}$, $Nd^{3+}$, $Sm^{3+}$, $Eu^{3+}$, $Ho^{3+}$, $Er^{3+}$, and $Yb^{3+}$ [12]. As the temperature increases, the relative intensity of light emitted from the higher energy level rises, forming the basis of the fluorescence intensity ratio (FIR) method (green box in **Figure 1**). By observing the ratio between two peaks in the emission spectrum, we can precisely determine temperature. This approach is straightforward, versatile, and less susceptible to experimental errors or misalignments, making it a preferred choice for accurate temperature sensing method. The only constraint is that energy levels they contain must be closely spaced [22].

The appeal of rare-earth-doped materials for temperature sensors has risen due to their economical fabrication and the ease of excitation using low-cost diode lasers. These materials contain multiple pairs of energy levels with small separations, comparable to thermal energy. In practical sensor applications, these energy levels are not only optically connected to the ground state but also have a high likelihood of non-radiative transitions between paired levels [23].

In the field of luminescence thermometry employing lanthanides, the pivotal component lies within the $Er^{3+}$ ion, with particular emphasis on its $^4S_{3/2}$ and $^2H_{11/2}$ levels, which are prominently employed. As an example, **Figure 2** exhibits a segment of the erbium emission spectrum in an $Er^{3+}$-doped YGG nano-garnet ($Er_{0.1}Ho_{0.1}$:$Y_{2.8}Ga_5O_{12}$).

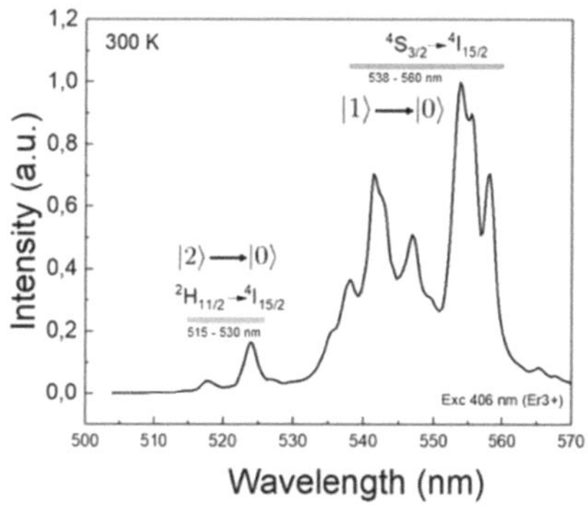


**Figure 2.**
*Emission spectrum of a sample (an $Er^{3+}$-doped YGG nano-garnet) at 300 K showcasing distinctive peaks corresponding to fluorescence emissions from erbium ions under the specified excitation condition.*

The 523-nm and 550-nm emission bands are attributed to transitions within $Er^{3+}$ ions, specifically from $^2H_{11/2}$ to $^4I_{15/2}$ and from $^4S_{3/2}$ to $^4I_{15/2}$ energy levels, respectively. A distinct advantage of FIR thermometry resides in its ability to easily discern the sharp line emissions originating from these two levels. This clarity significantly simplifies the identification process of the $^2H_{11/2}$ and $^4S_{3/2}$ levels, particularly when contrasted with scenarios where bands overlap, presenting challenges in their differentiation.

At typical room temperature, there exists an equilibrium state between the $^4S_{3/2}$ and $^2H_{11/2}$ levels, and both levels undergo light emission, with the $^4S_{3/2}$ level displaying a notably stronger emission compared to the $^2H_{11/2}$ level. These levels, referred to as thermal coupling levels (TCLs), demonstrate a discernible energy discrepancy ($E_{21}$) of approximately 900 $cm^{-1}$, as illustrated in **Figure 3**. Owing to the narrow gap separating these levels, their population distributions adhere closely to a Boltzmann distribution pattern as the temperature undergoes fluctuations.

As temperature rises, there is a notable increase in emissions from the $^2H_{11/2}$ level, as depicted in **Figure 4** [24]. The positions of the two bands remain constant while the temperature increases. This rise in intensity is crucial for FIR thermometry analysis. By comparing the intensities of these two emissions, precise delineation of the areas corresponding to the $^4H_{3/2}$ and $^2H_{11/2}$ emissions is facilitated. These areas will be used to obtain the FIR by the expression:

$$FIR = \frac{Area_{|2\rangle \to |0\rangle}}{Area_{|1\rangle \to |0\rangle}} \tag{3}$$

These delineated areas serve as the basis for determining the relative intensity, which in turn allows for the calculation of the temperature of the environment. By analyzing the ratio of their intensities, a calibration curve can be generated. The methodology for determining temperature relies on meticulously assessing the relative intensities of the $^4S_{3/2}$ and $^2H_{11/2}$ emissions. This comparative analysis allows us to precisely understand the temperature variations within the environment, aiding in practical applications, such as optimizing thermal management systems or ensuring proper operation of heating or cooling equipment. Therefore, the FIR technique provides an effective method to calibrate optical temperature sensors [15], improving measurement sensitivity and reducing the influence of varying measurement conditions [23].

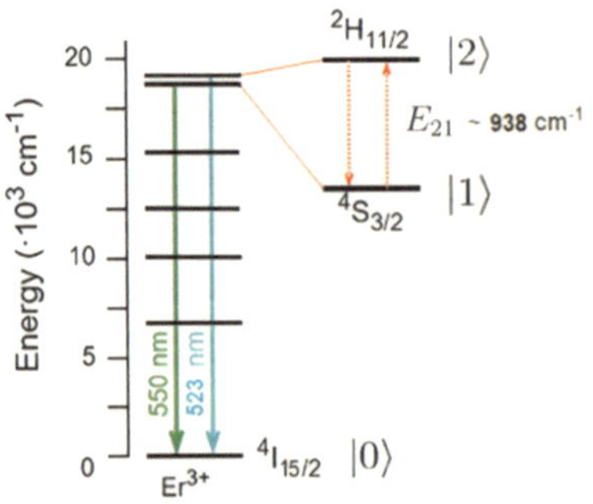


**Figure 3.**
*Diagrams illustrating the energy levels and transitions utilized in the FIR method, focusing on thermally coupled excited states of $Er^{3+}$ focusing on TCLs $^2H_{11/2}$ and $^4S_{3/2}$.*

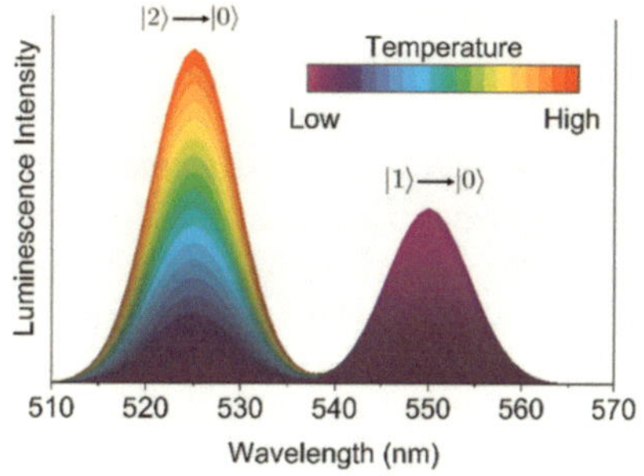


**Figure 4.**
*Diagram showing how emission bands in luminescent materials doped with erbium ions vary with temperature. Adapted from Ref. [24].*

## 2. Fundamentals

Lanthanide-doped materials hold a distinct position in photonics, thanks to the exceptional spectroscopic properties exhibited by $Ln^{3+}$ ions in terms of luminescence and amplification. The pivotal discovery of the bright red-emitting phosphor $Y_2O_3:Eu^{3+}$ at the dawn of the twentieth century marked a significant milestone in this field. Furthermore, the introduction of the $YAG:Nd^{3+}$ laser in 1964 further emphasized the crucial role of lanthanides in shaping modern photonic technologies. This is attributed to their remarkable photoluminescence properties that stem from their abundant energy level structure [25]. This rich energy level structure, depicted in the Dicke diagram reprinted in Ref. [25], highlights the intricate electronic configurations of lanthanides, contributing to their diverse and versatile photoluminescent capabilities and distinctive characteristics that set them apart. Primarily, they showcase exceptional spectroscopic properties. These include emitting sharp lines, maintaining stable energy levels, and absorbing light across a broad spectrum, from infrared to ultraviolet. Their electronic structure grants them insulation from external influences within a host environment, ensuring that their transition energies remain independent. This shielding effect preserves the positions of their energy levels, minimizing any alterations when incorporated into a host matrix. Certain ions exhibit luminescence in the visible or near-infrared spectral (NIR) regions when subjected to UV irradiation. The emitted light's color varies depending on the specific $Ln^{3+}$ ion present. For example, $Eu^{3+}$ emits red light, $Tb^{3+}$ emits green light, $Sm^{3+}$ emits orange light, and $Tm^{3+}$ emits blue light [26]. $Yb^{3+}$, $Nd^{3+}$, and $Er^{3+}$ ions are renowned for their NIR luminescence. Additionally, $Pr^{3+}$, $Sm^{3+}$, $Dy^{3+}$, $Ho^{3+}$, and $Tm^{3+}$ ions also feature transitions in the NIR region, while $Gd^{3+}$ emits exclusively in the ultraviolet (UV) region [25].

### 2.1 Energy levels in lanthanide ions

Lanthanides (Ln) encompass a series of elements in the periodic table spanning from cerium (Z = 58) to lutetium (Z = 71), succeeding lanthanum (Z = 57). They are characterized by the progressive filling of their 4f orbitals within their electronic configuration, ranging from $[Ar]5s^2\,5p^6\,4f^1\,6s^2$ in cerium to $[Ar]5s^2\,5p^6\,4f^{13}\,6s^2$ in ytterbium. In their +3-oxidation state (commonly found in solids as $Ln^{3+}$), lanthanides typically shed two electrons from the 6 s orbital and one from the 4f orbital, leaving the 4f orbital partially filled. The electronic configuration of the $Ln^{3+}$ ions is represented by $[Xe]4f^N$ [25, 27].

The intense shielding of the 4f electrons by the complete $5s^2$, $5p^6$, and $6s^2$ shells is a defining trait of lanthanide ions, rendering them behaviorally akin to atoms. While their energy levels exhibit stability transitioning from a free state to a solid matrix, their spectroscopic characteristics, such as absorption and emission band widths, are influenced by the host matrix's nature. The arrangement of neighboring ions around the lanthanide ion holds significance in various matrices, particularly in crystalline versus disordered structures, where the former yields narrow bands due to uniform crystal field effects, while the latter leads to widened bands owing to varied ligand field energy levels. Heterogeneous broadening, characteristic of amorphous materials like glasses, arises from the lack of periodicity and variation in the bonding environment. This variability affects the Stark sublevels, resulting in broad absorption bands. This phenomenon is detailed in studies on glasses and luminescent materials doped with rare-earth ions, where it is observed that the absence of periodicity in glasses leads to variation in spectroscopic properties [28].

The shielding provided by the filled 5s and 5d shells, along with the partially filled $6s^2$ shell, insulates lanthanide ions against external fields from the host medium, thereby preserving the spectral characteristics of emission lines and minimizing distortions, akin to free ions.

The trivalent state of lanthanides holds several unique characteristics. These characteristics encompass distinct spectroscopic properties, exemplified by sharp lines observed in the $Er^{3+}$ emission spectrum of **Figure 2**. Moreover, the independence of energy levels from the host environment is illustrated in **Figure 5**, where the absorption spectra of $Ho^{3+}$ ions in two distinct hosts display identical configurations.

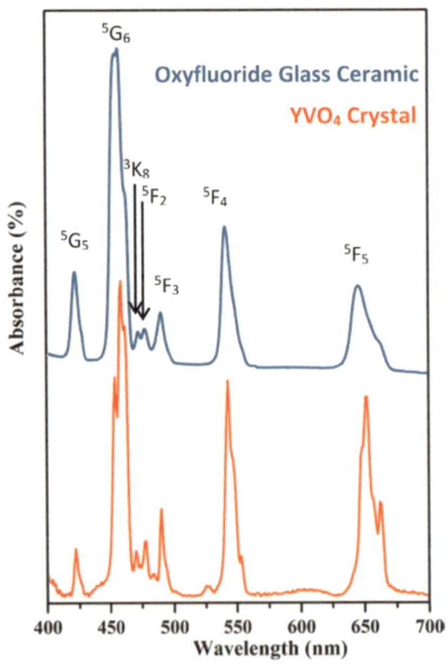


**Figure 5.**
*Absorption spectra of an oxyfluoride glass-ceramic and yttrium orthovanadate ($YVO_4$) crystal doped with erbium ions highlight the independence of energy levels from the host environment.*

Additional, lanthanide ions exhibit emissions ranging from infrared to ultraviolet, as shown in the Dicke diagram [25]. Trivalent lanthanides also exhibit long emission lifetimes and high quantum yields [25].

Energy from an electronically excited state dissipates through both radiative and non-radiative means [20]. The emitted intensity correlates with the electron population density (N) in excited states, and its temporal changes are governed by

$$\frac{dN}{dt} = -(k_R + k_{NR})N(t) \tag{4}$$

The rates of radiative ($k_R$) and non-radiative ($k_{NR}$) transitions determine the processes of photon emission. Radiative de-excitation pathways lead to photon emission, while non-radiative pathways release vibrational energy. Consequently, the electron population in the excited state and the luminescence intensity decay exponentially over time, with a characteristic time constant ($\tau$) often referred to as the lifetime of the excited state.

$$\tau = \frac{1}{k_R + k_{NR}} \tag{5}$$

The reciprocal of the radiative transition rate is known as the radiative lifetime or natural lifetime, denoted as $\tau_R$:

$$\tau_R = \frac{1}{k_R} \tag{6}$$

This quantity can often be accurately calculated from absorption and emission spectra, as well as from the ratio of the measured lifetime to the internal quantum efficiency of emission, also known as the quantum yield ($\eta$):

$$\tau_R = \frac{\tau}{\eta} \tag{7}$$

Transition metals are well known for their characteristic d-orbital electronic transitions, which result in complex and varied spectra. Similarly, lanthanides possess distinct f-orbital electronic configurations that lead to their own set of spectral features. These differences lie once more in their distinctive electronic configuration. The 4f electrons of lanthanides are shielded by the 5s and 5p electrons. As a result, when trivalent lanthanides are embedded within a host matrix, the influence of the host ions is typically minimal [25].

To explore the complexities of the 4f configuration, $Eu^{3+}$ serves as an illustrative example (**Figure 6**). The electronic configuration of $Eu^{3+}$ is [Xe] $4f^6$. Initially, interactions between the core and the electrons yield a $4F^6$ configuration, with europium accommodating six electrons in its 4f shell. Subsequent introduction of electron repulsion results in terms displaying significant splitting into J-levels, further accentuated by spin-orbit interaction, which segregates these terms into distinct levels. These properties are indicated by term symbols S, L, and J in the notation $^{2S+1}L_J$. Here, 2S + 1 denotes the spin multiplicity, L represents the total orbital angular momentum, and J signifies the total angular momentum of the 4f electrons [25].

Because of the screening effect of 4f electrons, their interaction with the crystal field of ligands is significantly weaker than their spin-orbit interaction. The crystal

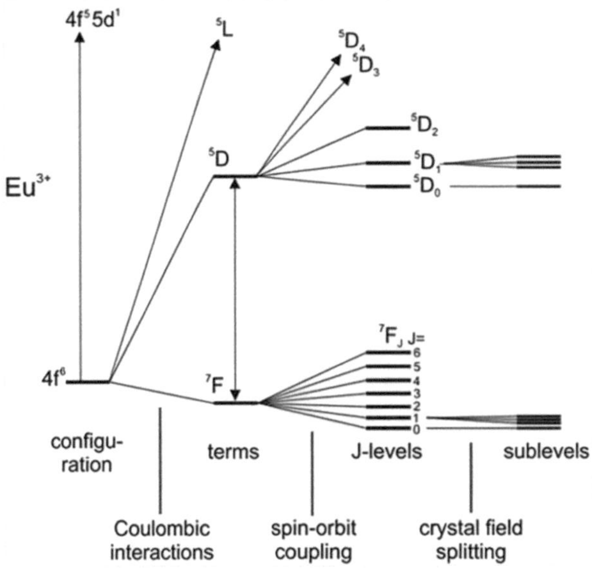


**Figure 6.**
*Diagram illustrating the interactions causing the splitting of electronic energy levels of a $Eu^{3+}$ ion. Energy increases as depicted upward in the diagram [29].*

field causes only a slight change in energy, resulting in the splitting of levels into multiple sublevels, known as the Stark effect. The maximum number of these Stark sublevels is 2J + 1 [20]. However, the splitting and the specific number of sublevels are influenced by the symmetry of the crystal field surrounding the $Ln^{3+}$ ion and the corresponding selection rules [20].

The selection rules for electric dipole transitions are derived from the Judd-Ofelt (JO) theory. This theory provides a framework for understanding and predicting the intensity of spectral lines in lanthanide ions by considering the interaction between electronic states and the surrounding crystal field. It accounts for the otherwise forbidden transitions by introducing mixed parity states, which allow for electric dipole transitions under non-centrosymmetric conditions. Consequently, the Judd-Ofelt (JO) theory not only explains the occurrence of these transitions but also establishes the criteria for their probabilities based on the symmetry and environment of the lanthanide ions.

The trivalent lanthanides can be treated as free ions, where the Hamiltonian accounts for their behavior. The surrounding crystal field perturbs the system, contingent upon crystal field parameters, inducing additional splitting. The J-levels can be further split into sublevels because of the electric field of the matrix. However, due to shielding, energy level positions weakly hinge on the host matrix, rendering transition energies predominantly independent, as evidenced by the Dicke diagram [29].

Within the host matrix context, level splitting varies based on symmetry considerations, often yielding diverse peak numbers. However, comprehending intensity origins remains a challenge in lanthanide ion studies within the host matrix.

The metastable energy bands in $Ln^{3+}$ ions, influenced by factors like electronic repulsion and spin-orbit coupling, contribute to the structured arrangement of energy levels within these ions. This structured arrangement is crucial for understanding the spectroscopic properties of lanthanide ions, particularly in materials like glasses and luminescent materials doped with rare-earth ions. The JO theory, which provides a framework for analyzing the spectral properties of lanthanide ions, builds upon our understanding of these structured energy levels. By elucidating the energies of various multiplets and predicting important properties related to light emission, the JO theory enhances our comprehension of optical materials, bridging the gap between fundamental energy level organization and practical spectroscopic analysis. Initially, in the early twentieth century, scientists struggled to comprehend the spectral properties of these ions. Although they recognized the sharp lines in their spectra, explaining why they appeared was challenging due to their violation of certain rules in quantum mechanics, like the Laporte rule.

In the 1940s, Raka developed his renowned algebra, laying the groundwork for complex calculations facilitated by the advent of computers. Soon thereafter, a breakthrough emerged: within non-centrosymmetric crystal fields, coupling between odd and even parity states forced the creation of mixed parity states, effectively mitigating the Laporte rule. So, in 1962, Judd and Ofelt made significant contributions by solving the Schrödinger equation for the steady state of a many-electron system. This breakthrough enabled them to successfully ascertain the energies of the respective $^{2S+1}L_J$ multiplets, which are pivotal for comprehending the emission of light by $Ln^{3+}$ ions. Additionally, their JO theory pioneered a fresh approach to forecast crucial properties relevant to light emission analysis, marking a noteworthy advancement in the field of spectroscopy [30]. Despite its inherent complexity, the JO theory enables the prediction of crucial derived quantities, such as transition probabilities, branching ratios, lifetimes, and quantum efficiencies, utilizing just three parameters ($\Omega_1$, $\Omega_2$, and $\Omega_3$). These parameters provide relevant information on the local environment of the rare-earth ions [31], making JO theory a powerful tool in understanding and optimizing optical materials.

### 2.2 Boltzmann equilibrium

When analyzing $Ln^{3+}$-doped materials, the FIR technique quantifies the emission intensities originating from closely spaced excited states (also known as the Boltzmann-type FIR method). It also evaluates intensities from electron transitions ending at various Stark sublevels. This method is versatile, suitable for both down-conversion and up-conversion emissions [20, 21]. By analyzing the fluorescence intensities of two adjacent energy levels that are in thermal equilibrium and optically coupled to a common lower level, temperature variations can be determined. These levels are thermally coupled due to their high probability of non-radiative transitions. The FIRs, which follow the Boltzmann distribution law, are calculated from the emission peaks of these thermally coupled energy levels at different temperatures. Two excited energy states of lanthanide ions are termed thermally coupled if the energy gap between them is 2000 $cm^{-1}$ or less. This proximity enables electrons to transition between these states due to thermal energy. Consequently, both states share the electronic population following Boltzmann's distribution:

$$N_2 = N_1 exp\left(-\frac{E_{21}}{kT}\right) \tag{8}$$

$N_2$ and $N_1$ represent the numbers of electrons in the higher and lower excited states, respectively. $E_{21}$ denotes the energy difference between these states. The absolute temperature (T) determines how electrons distribute between these states according to Boltzmann's distribution. This distribution explains the probability of electrons occupying different energy levels based on the thermal energy available at temperature T. Consequently, the fluorescence intensity ratio (FIR) emitted from the higher ($I_{20}$) and lower ($I_{10}$) excited states can be estimated as follows:

$$FIR = \frac{I_{20}}{I_{10}} = \frac{\omega_{20}^{R} g_2 h\nu_2}{\omega_{10}^{R} g_1 h\nu_1} exp\left(-\frac{E_{21}}{kT}\right) = B\ exp\left(-\frac{E_{21}}{kT}\right) \tag{9}$$

Here, $h$ represents Planck's constant, $g$ signifies the degeneracy (2J + 1) of the energy states, A represents the rate of spontaneous emission, $\nu$ denotes the emission frequency, 2 and 1 refer to the "higher" and "lower" energy states, respectively, 0 denotes the ground state or an electronic level with lower energy than 2 and 1, and $\omega_{20}^{R}$ and $\omega_{10}^{R}$ are the spontaneous emission rates of the $E_2$ and $E_1$ levels to the $E_0$ level, respectively.

The log(FIR) exhibits a linear relationship with the reciprocal of temperature:

$$\log(FIR) = \log(B) - \frac{E_{21}}{k}\frac{1}{T} \tag{10}$$

Thus, the values of log(B) and $E_{21}/k$ can be experimentally determined from the slope and intercept, respectively, of the linear fit of log (FIR) versus 1/T.

At colder temperatures, the higher energy state often remains unoccupied due to electrons lacking sufficient energy to bridge the energy gap. Moreover, when energy states are closely spaced, the non-radiative relaxation rate from the higher to the lower energy state becomes significantly pronounced. Consequently, the FIR read-out method has a temperature threshold proportional to $E_{21}$. This implies that lower values of $E_{21}$ correspond to colder temperatures suitable for FIR measurements. As temperature rises, the higher energy state sees an increasing population, leading to a gradual rise in emission intensity from this state at the expense of reducing the population in the lower energy state. As temperature rises, both emissions from the higher and lower energy states decrease in intensity due to a phenomenon known as temperature quenching. This reduction in intensity continues until one or both emissions become so weak that they are no longer detectable. The upper temperature limit for applying the FIR method depends largely on the phonon spectrum of the material hosting the luminescent material and the specific lanthanide ion being used. Essentially, the ability to accurately measure temperature diminishes as temperatures approach these upper limits, mainly due to the increased uncertainties associated with detecting weakened emissions. The FIR method allows for the use of any emissions if they originate from two excited levels that are thermally coupled. This flexibility enables the method to be applied to a wide range of materials and experimental conditions, thereby enhancing its versatility in temperature sensing applications.

## 3. The performance of a luminescence thermometer

To evaluate the effectiveness of thermometers, various parameters come into play [20]. One key parameter is absolute sensitivity ($S_{FIR}$), which quantifies the change in FIR per temperature increment. However, perhaps more commonly considered is relative sensitivity ($S_R$), which measures the change in FIR per temperature increment relative to that FIR itself.

Both $S_{FIR}$ and $S_R$ can be derived from Eq. (9) as follows:

$$S_{FIR} = \frac{\partial(FIR)}{\partial T} = \frac{E_{21}}{kT^2}(FIR) \tag{11}$$

$$S_R = \frac{\partial(FIR)}{\partial T}\frac{1}{(FIR)} = \frac{E_{21}}{kT^2} \tag{12}$$

Maximizing these parameters is crucial for optimal thermometer performance. Even a slight temperature fluctuation can enhance sensitivity, indicative of a superior thermometer's ability to accurately detect temperature changes. Another important metric in assessing thermometer performance is the limit of detection, denoted as $\delta T_{min}$. This metric is closely linked to the minimum temperature change that the thermometer is capable of registering. In practical terms, $\delta T_{min}$ represents the smallest temperature variation that the thermometer can reliably detect and measure within its operational range. It's mathematically defined as follows:

$$\delta T_{min} = \frac{\delta(FIR)}{(FIR)}\frac{1}{S_R} \tag{13}$$

The term $\delta(FIR)/(FIR)$ represents the resolution limit or relative uncertainty of the FIR, indicating the smallest detectable change in the fluorescence intensity ratio. Enhancing this limit may require improving equipment and refining measurement techniques. For example, increasing integration time and averaging measurements can reduce experimental noise, thus improving resolution [16]. The uncertainty of the FIR is expressed as follows:

$$\frac{\delta(FIR)}{(FIR)} = \sqrt{\left(\frac{\delta Area_{|1\rangle\to|0\rangle}}{Area_{|1\rangle\to|0\rangle}}\right)^2 + \left(\frac{\delta Area_{|2\rangle\to|0\rangle}}{Area_{|2\rangle\to|0\rangle}}\right)^2} \tag{14}$$

where $Area_{|1\rangle\to|0\rangle}$ and $Area_{|2\rangle\to|0\rangle}$ represent the areas associated with the emissions used to calibrate the thermometer and $\delta Area_{|1\rangle\to|0\rangle}$ and $\delta Area_{|2\rangle\to|0\rangle}$ are their uncertainties, measured as the standard deviations of these areas from multiple measurements taken under uniform conditions [16].

In a recent investigation [16], the sensitivity of the sensor to temperature variations was explored using both $Er^{3+}$ and $Ho^{3+}$ ions co-doped in nano-garnet as the emissive species. The FIR technique was employed for temperature detection across a broad range, from 30 K to 540 K. The thermally coupled levels $^2H_{11/2}$ and $^4S_{3/2}$ for $Er^{3+}$ ions and $^5S_2$ and $^5F_4$ for $Ho^{3+}$ ions were utilized in the temperature sensing mechanism. This approach is significant, as few optical sensors are capable of effectively covering both high and low temperature ranges.

Upon excitation of the sample with lasers, distinct emission bands corresponding to both $Er^{3+}$ and $Ho^{3+}$ ions were observed, as shown in **Figure** 7. The emission bands facilitated accurate temperature measurements through the FIR technique, demonstrating the sensor's efficacy in detecting temperature variations over a wide range, thereby offering potential advancements in optical temperature sensing technologies.

Allowing for independent excitation of $Er^{3+}$ or $Ho^{3+}$ ions facilitates the calculation of $S_R$ for emissions from both ions (**Figure 8**).

For $Er^{3+}$ ions, the $S_R$ was observed to increase notably from approximately 0.2% $K^{-1}$ at 540 K to 1.3% $K^{-1}$ at 200 K [16]. This dynamic underscores the sensor's

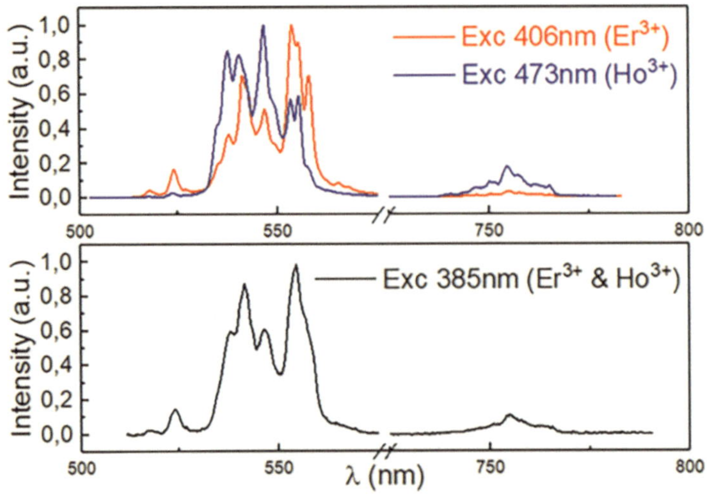


**Figure 7.**
*Spectrum showing emissions from the sample when simultaneously excited at 385 nm and independently excited at 406 nm for $Er^{3+}$ ions, and at 473 nm for $Ho^{3+}$ ions. Adapted from Ref. [16].*

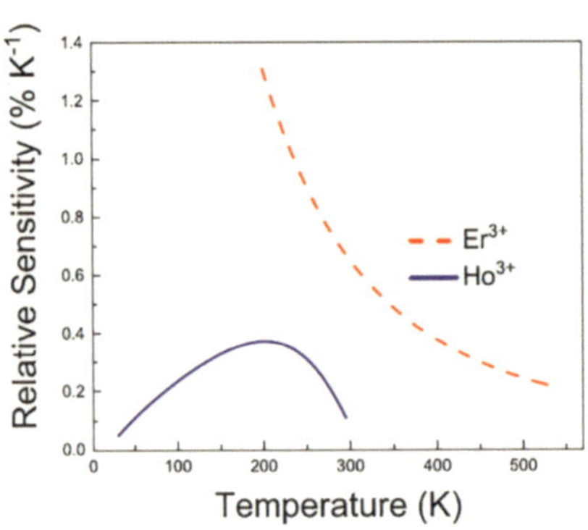


**Figure 8.**
*Relative sensitivity of the sensor for YGG nano-garnet, measured independently for erbium (red, dashed line) and holmium (blue, solid line) transitions [16].*

heightened responsiveness at lower temperatures, where even minor temperature changes prompt significant variations in the fluorescence intensity ratio. Conversely, at higher temperatures such as 540 K, the sensor exhibits a comparatively muted response to temperature fluctuations, reflecting reduced sensitivity. Meanwhile, the evaluation of $S_R$ for $Ho^{3+}$ ions across temperatures ranging from 30 K to 300 K revealed a peak sensitivity of about 0.4% $K^{-1}$ at 200 K [16]. This finding indicates effective temperature sensing capabilities at lower temperature regimes for $Ho^{3+}$ ions, albeit with sensitivity characteristics distinct from those observed for $Er^{3+}$ ions.

## 4. Conclusion

Optical FIR-based thermometry emerges as a cutting-edge technology poised to revolutionize temperature measurement across various disciplines. By leveraging the fluorescence properties of lanthanide ions, particularly erbium and holmium in doped nano-garnets, researchers have advanced our understanding of temperature-sensitive materials. The technology offers non-invasive and highly sensitive methods capable of operating from cryogenic to high temperatures, making it invaluable in diverse applications from materials science to biomedical research. Challenges remain in enhancing resolution and reducing experimental noise, yet ongoing advancements in material design and measurement techniques promise continued innovation. Optical FIR-based thermometry stands at the forefront of temperature sensing technologies, offering profound implications for future scientific and industrial applications.

## Acknowledgements

The author wishes to thank the Direção Regional da Ciência, Inovação e Desenvolvimento, Governo Regional dos Açores, for their generous support through application M3.3.C/EDIÇÕES/022/2024 – "Fundamental Concerns of Optical FIR-Based Thermometry".

## Author details

Helena Cristina Vasconcelos[1,2]

1 Faculty of Sciences and Technology, Azores University, Ponta Delgada, Açores, Portugal

2 Laboratory of Instrumentation, Biomedical Engineering and Radiation Physics (LIBPhys-UNL), Department of Physics, NOVA School of Science and Technology, NOVA University Lisbon, Almada, Portugal

*Address all correspondence to: helena.cs.vasconcelos@uac.pt

## References


[1] McGee TD. Principles and Methods of Temperature Measurement. New York: John Wiley and Sons, Inc.; 1988

[2] Soulen RJ. A brief history of the development of temperature scales: The contributions of Fahrenheit and Kelvin. Superconductor Science and Technology. 1991;**4**:696-699

[3] Wright WF. Early evolution of the thermometer and application to clinical medicine. Journal of Thermal Biology. 2016;**56**:18-30. DOI: 10.1016/j.jtherbio.2015.12.0

[4] Chang H, Yi SW. The absolute and its measurement: William Thomson on temperature. Annals of Science. 2005; **62**(3):281-308. DOI: 10.1080/00033790410001712246

[5] Young HD, Freedman RA, Ford AL. University Physics with Modern Physics. 15th ed, illustrated, reprint. Boston, MA: Pearson; 2019

[6] Michalski L, Eckersdorf K, Kucharski J, McGhee J. Temperature Measurement. 2nd ed. West Sussex, UK: John Wiley and Sons LTD; 2001

[7] Ryszczyńska S, Martín IR, Grzyb T. Near-infrared optical nanothermometry via upconversion of Ho3+−sensitized nanoparticles. Scientific Reports. 2023;**13**: 14819. DOI: 10.1038/s41598-023-42034-z

[8] Rai VK. Temperature sensors and optical sensors. Applied Physics. B. 2007; **88**(2):297-303

[9] Xiang G, Xia Q, Liu X, Wang X. Optical thermometry based on the thermally coupled energy levels of Er3+ in upconversion materials. Dalton Transactions. 2020;**49**(47):17115-17120. DOI: 10.1039/d0dt03100c

[10] Jaque D, Vetrone F. Luminescence nanothermometry. Nanoscale. 2012; **4**(15):4301. DOI: 10.1039/c2nr30764b

[11] Sole JG, Bausa LE, Jaque D. An Introduction to the Optical Spectroscopy of Inorganic Solids. West Sussex: John Wiley and Sons; 2005

[12] Dagupati R, Klement R, Galusek D. Er 3+/Yb 3+ co-doped oxyfluoro tellurite glasses: Analysis of optical temperature sensing based on up-conversion luminescence. International Journal of Applied Glass Science. 2021;**12**(4): 462-471. DOI: 10.1111/ijag.16011

[13] Brites CDS, Millán A, Carlos LD. Lanthanides in luminescent thermometry. Handbook on the Physics and Chemistry of Rare Earths. 2016;**49**: 339-427. DOI: 10.1016/bs.hpcre.2016.03.005

[14] Zhou Y, Qin F, Zheng Y, Zhang Z, Cao W. Fluorescence intensity ratio method for temperature sensing. Optics Letters. 2015;**40**:4544-4547

[15] Zhou S, Li X, Wei X, Duan C, Yin M. A new mechanism for temperature sensing based on the thermal population of 7F2 state in Eu3+. Sensors and Actuators B: Chemical. 2016;**231**: 641-645. DOI: 10.1016/j.snb.2016.03.082

[16] Soler-Carracedo K, Martín IR, Lahoz F, Vasconcelos HC, Lozano-Gorrín AD, Martín LL, et al. Er3+/Ho3+ codoped nanogarnet as an optical FIR based thermometer for a wide range of high and low temperatures. Journal of Alloys and Compounds. 2020:156541. DOI: 10.1016/j.jallcom.2020.156541

[17] Berthou H, Jorgensen CK. Optics Letters. 1990;**15**:1100

[18] Kumar R, Binetti L, Nguyen TH, Alwis LSM, Sun T, Grattan KTV. Optical fibre thermometry using ratiometric green emission of an upconverting nanoparticle-polydimethylsiloxane composite. Sensors and Actuators A: Physical. 2020;**312**:112083. DOI: 10.1016/j.sna.2020.112083

[19] Lozano-Gorrín AD, Rodríguez-Mendoza UR, Venkatramu V, Monteseguro V, Hernández-Rodríguez MA, Martín IR, et al. Lanthanide-doped Y3Ga5O12 garnets for nanoheating and nanothermometry in the first biological window. Optical Materials. 2018;**84**:46-51. DOI: 10.1016/j.optmat.2018.06.043

[20] Dramićanin M. Luminescence Thermometry: Methods, Materials, and Applications. Elsevier; 2018. pp. 1-292. DOI: 10.1016/C2016-0-01905-8

[21] Martins JC, Brites CDS, Neto ANC, Ferreira RAS, Carlos LD. An overview of luminescent primary thermometers. In: Martí JJC, Baiges MCP, editors. Luminescent Thermometry: Applications and Uses. Cham: Springer; 2023. DOI: 10.1007/978-3-031-28516-5_3

[22] Wade SA, Collins SF, Baxter GW. Fluorescence intensity ratio technique for optical fiber point temperature sensing. Journal of Applied Physics. 2003;**94**(8):4743-4756

[23] Haro-González P, León-Luis SF, González-Pérez S, Martín IR. Analysis of Er3+ and Ho3+ codoped fluoroindate glasses as wide range temperature sensor. Materials Research Bulletin. 2011;**46**(7):1051-1054. DOI: 10.1016/j.materresbull.2011.0

[24] Hernández-Álvarez C, Brito-Santos G, Martín IR, Sanchiz J, Saidi K, Soler-Carracedo K, et al. Journal of Materials Chemistry C. 2023;**11**: 10221-10229

[25] Khan LU, Khan ZU. Rare earth luminescence: Electronic spectroscopy and applications. In: Sharma S, editor. Handbook of Materials Characterization. Cham: Springer; 2018. pp. 345-404. DOI: 10.1007/978-3-319-92955-2_10

[26] Binnemans K. Lanthanide-based luminescent hybrid materials. Chemical Reviews. 2009;**109**:4283-4374

[27] Khan LU, Khan ZU. Bifunctional nanomaterials: Magnetism, luminescence and multimodal biomedical applications. In: Complex Magnetic Nanostructures. Cham: Springer International Publishing; 2017. pp. 153-156

[28] Molières E, Panczer G, Guyot Y, Jollivet P, Majérus O, Aschehoug P, et al. Investigation of local environment around rare earths (La and Eu) by fluorescence line narrowing during borosilicate glass alteration. Journal of Luminescence. 2014;**145**:213-218. DOI: 10.1016/j.jlumin.2013.07.051

[29] Vuojola J, Soukka T. Luminescent lanthanide reporters: New concepts for use in bioanalytical applications. Methods and Applications in Fluorescence. 2014;**2**(1):012001. DOI: 10.1088/2050-6120/2/1/012001

[30] Hehlen MP, Brik MG, Krämer KW. 50th anniversary of the Judd–Ofelt theory: An experimentalist's view of the formalism and its application. Journal of Luminescence. 2013;**136**:221-239. DOI: 10.1016/j.jlumin.2012.10.035

[31] Cantelar E, Sanz-García JA, Sanz-Martín A, Muñoz Santiuste JE, Cussó F. Structural, photoluminescent properties and Judd-Ofelt analysis of Eu3+-activated CaF2 nanocubes. Journal of Alloys and Compounds. 2020;**813**: 152194. DOI: 10.1016/j.jallcom.2019.152194

Chapter 3

# Perovskite Quantum Dots: A New Generation of Promising Scintillator Materials

*Daihao Li, Tianming Sun, Stephen Kearney, Robert Moss and Mingqing Wang*

## Abstract

Scintillator materials are essential parts of X-ray imaging/detection for medical diagnostics, non-destructive detection, security inspection, and space exploration. Traditional commercial scintillators are limited by low luminescence efficiency and afterglow effect. Perovskite ($ABX_3$) quantum dots (PQDs), with the ability to convert X-ray radiation into multi-color visible light, has emerged as a new class of competitive scintillators, based on the advantages of high X-ray absorption coefficients, fast luminescence, high luminescence efficiency, and low-cost wet chemical facile synthesis on various substrates. Understanding the relationship between the structure and luminescence of PQDs is essential for scintillator discovery and optimization. This chapter first reviewed the work principle, materials selection, and key parameters of indirect X-ray detectors, followed by the basic crystal structure, optoelectronic properties, and synthesis methods for PQDs. Then the recent advances in perovskite scintillators are comprehensively reviewed. Finally, we summarize the current challenge in perovskite scintillators with promising solutions and provide a perspective on the future direction of this emerging scintillator.

**Keywords:** perovskite quantum dots, scintillator, X-ray indirect detector, X-ray detector, X-ray imaging

## 1. Introduction

X-Ray radiation, uncovered by Wilhelm Röntgen circa 1895, comprises a genre of high-intensity electromagnetic radiation with extraordinary penetrative properties, offering provisions for non-destructive inspection [1]. Operating in medical radiography, photodynamic therapy (PDT), nuclear deterrence, and security audits, X-rays have become an indispensable tool [2]. The operative principle of X-ray detection hinges on tracking and disclosing the post-penetration diminishing of incident X-ray radiation on specific subjects. For instance, the inspection of in-vivo organs renders high-resolution imaging instrumental in clinical diagnostics, as does the exploration of non-living matter for industrial and security purposes [3].

The current X-ray detection machinations are bifurcated into two central categories: direct conversion, where X-ray quantum is translated to an electrical signal via a semiconductor, or to a chemical signal via a film, and indirect conversion, where the transformation is to low-energy photons (visible light) mediated by scintillating materials [4, 5]. Presently, for X-ray imaging based on integrating detectors, there are limited materials that can replace nano-scintillators in the consumer market, outside of the selenium (Se) detector purposed for mammography. Even leading candidates for direct conversion materials, such as CdTe and CdZnTe (CZT), wrestle with issues such as a delay in response rate, coupled with temporal maladies like persistent afterglow incited by electron entrapment. Consequently, manufacturing large-sized wafers with diminished noise current yet high mobility-lifetime product ($\mu\tau$) remains an arduous task, rendering the achievement of an acceptable price-performance ratio for commercial applications a challenge (**Figure 1**).

Conversely, scintillators pave the way to convert X-rays into visible light, known as down-conversion, engendering the commercial utilization of affordable sensing arrays such as photomultiplier tubes (PMT), avalanche photodiodes (APD), amorphous-Si photodiode matrices, and charge-coupled devices. This manipulation has generated a surge of research interest toward indirect conversion scintillating materials for X-ray detection.

However, established scintillators such as NaI:Tl, CsI:Tl, $CdWO_4$ (CWO), $Bi_4Ge_3O_{12}$ (BGO), $Lu_2SiO_5$ (LSO), and $Lu_{2(1-X)}Y_{2X}SiO_5$ (LYSO) necessitate further enhancements. Despite their potent stopping power and heightened light emission, qualifying them for commercial application in medical imaging devices, there is a demand to streamline the manufacturing processes, circumvent persistent afterglow, and discover novel electron-transition energy levels for adaptable scintillation. Particularly in diagnostic radiology, the attainment of superior spatial resolution while minimizing radiation exposure is a critical priority.

Thus, lead-halide perovskite nanocrystals-based X-ray sensors are an emerging technology projected to have considerable application prospects in future radiography, thanks to their accelerated fabrication, rapid response, and elevated spatial resolution. Such halide perovskites typically present unique electronic characteristics, including a diminished trap density, heightened charge-carrier mobility, and an extended minority-carrier diffusion length, specifically in nanocrystal morphology.

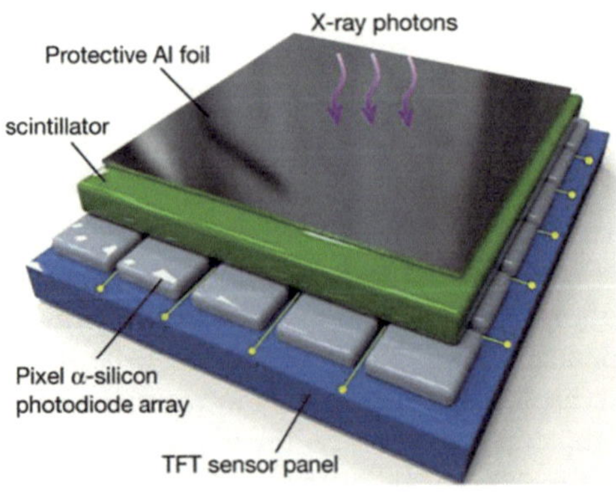


**Figure 1.**
*Application of scintillator in X-ray detectors [6].*

$CsPbCl_3$, despite demonstrating robust luminescence via free excitons upon exposure to radiation with prompt nanosecond response, was marred by issues such as intense self-absorption, decreased photoluminescence quantum yield (PLQY), Pb toxicity, and overall fragility. However, following decades of research, efficacious Pb-free scintillators, like double-perovskite, Cu, and Bi-based all-inorganic halide perovskites, have shown encouraging scintillation behaviors.

This review outlines the fundamentals of scintillators, their quintessential luminescence mechanisms, primary parameter performance, and the essential considerations for selecting optimal scintillator materials. A robust comparison is made between emerging perovskite quantum dot scintillators and traditional variants, emphasizing the merits, demerits, performance, and applications inherent to perovskite quantum dot systems. Finally, the obstacles encountered by perovskite quantum dots as a formidable contender in the realm of scintillator materials are stressed.

## 2. Fundamentals and progress of scintillators

### 2.1 Scintillators and their luminescence mechanism

Scintillators are defined as materials capable of emitting ultraviolet or visible light during the de-excitation phase having undergone stimulation by high-energy photons or irradiated particles. Presently, a broad array of inorganic scintillators is available, each embodying a distinct property. Concurrently, for Computed Tomography (CT) applications, it is imperative that the corresponding inorganic scintillators possess elevated density, augmented light yield, and minimal detection limits.

The process of scintillator luminescence is considerably intricate [7, 8], an aspect that has piqued extensive scientific curiosity, leading to persistent advancements. During the interactions between high-energy photons and the electrons within the lattice configuration, the conduction band experiences excitation, eliciting the integration of high-energy electrons and concurrently leaving hole states within the valence band. Upon surpassing the ionization threshold, the electron's energy provokes collision ionization, occasioning the generation of secondary electrons. If the energy is inadequate to foster ionization, the electrons and holes will implement energy reduction via lattice relaxation until they respectively migrate to the base of the conduction band and the apex of the valence band. Consequently, these can be bound to form excitons or traverse to the luminescent center to facilitate recombination. Explicitly, the luminescence process in scintillators can be classified into three operative phases: conversion, transfer, and lu-minescence, [9–11] as shown in **Figure 2**.

### 2.2 Criteria for selecting scintillator materials

An ideal scintillator will have essential features such as a high light yield, acute energy resolution, and swift time resolution. However, the reality demands that the chosen scintillator must cater to specific application requirements. In the context of CT imaging, the scintillator should exhibit strong X-ray absorption, minimal self-absorption, excellent light yield, and a short lifetime, optimally paired with the capacity for large-scale, arrayed production. This combination allows for the

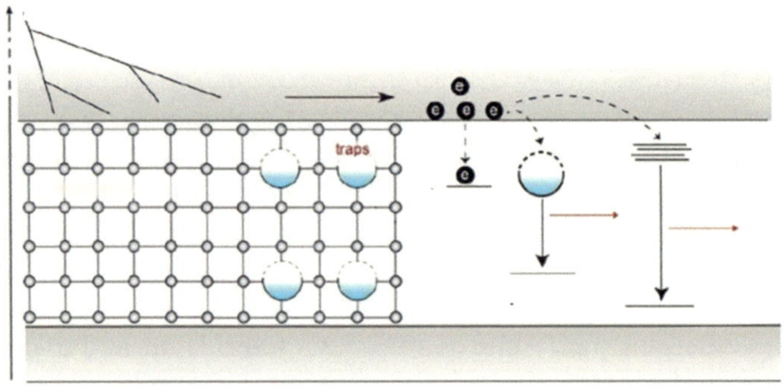


**Figure 2.**
*The luminescence mechanism of an inorganic scintillator is divided into three stages: conversion, transfer, and luminescence [11].*

realization of superior quality imaging under low-dosage conditions. The probable principles for selecting scintillators can be tabulated as follows:

1. Potent absorption capability: a larger quantity of incident X-ray photons absorbed by a scintillator translates to a higher conversion of hot electron-holes. X-ray absorption coefficient depends on the material's effective atomic number ($Z_{eff}$) and the density (ρ), hence, materials with a higher density and heavy elemental composition are deemed desirable.

2. Substantial light yield: light yield (LY), indicative of the scintillator's luminous efficacy, is quantified as the number of photons emitted per MeV of ionizing radiation [12]:

$$LY = \frac{1000000}{\beta E_g} SQ. \quad (1)$$

Where $E_g$ symbolizes the bandgap, β is a constant range between 2 and 3, S represents the transfer efficiency in the transfer process, and Q is the luminescence efficiency. As per its definition, light yield hinges on the bandgap $E_g$, the quality of the crystal, and the photoluminescence quantum efficiency (PLQY). Aiming for a high light yield mandates the chosen scintillator to possess a relatively narrow $E_g$ value, superior crystal quality, and a PLQY value as high as conceivable. An enhanced light yield is also beneficial for attaining a lower detection limit and a broader linear response, thereby improving the sensing efficiency.

3. Trivial self-absorption: self-absorption denotes a phenomenon where the scintillation photons are reabsorbed by the crystal—preventing their emission. Reducing self-absorption is a significant challenge in the field of scintillators. Minimized self-absorption implies a larger Stokes shift, and self-trapped exciton luminescent scintillators in this scenario possess an innate superiority.

4. Radioluminescence wavelength: the emission wavelength of a scintillator determines the photodetection coupling devices that can be selected. Ensuring minimal loss in collecting scintillation photons requires that the spectral detection range of the photodetector should overlap the emission range of the scintillation light.

5. Rapid decay: a faster decay lifetime signifies a sharp photoelectric conversion speed, imposing elevated time resolution of the scintillator detector. However, faster decay of light emission, at times, becomes unfavorable in obtaining a higher light yield. For CT imaging applications, a decay time (t) of less than 0.1 ms should suffice for this application requisites.

6. High stability: the stability of a scintillator pegs its usable life and serves as a prerequisite for its application. It mainly encompasses physicochemical stability, luminescent stability, and irradiation stability, which would accordingly impact the structural stability of the scintillator, its light yield, and irradiation hardness parameters.

7. Other aspects, such as low detection limit, broader linear response, economical, weak background radiation, upscaling potential, suitability for arrayed production, and more. Lower detection limits imply enhanced sensitivity of the detector for low-dose X-rays. Also, cost efficiency remains a significant consideration for practical applications.

Image quality parameters include spatial resolution, presenting the image's contrast and clarity within a specific spatial frequency range, generally quantified by the modulation transfer function (MTF) deploying the edge method. Image quality is generally regulated by the intrinsic properties of the scintillator, the detector, and their coupling. We will principally discuss the scintillator's influence.

## 3. Perovskite as promising scintillator materials

Perovskite is a promising material that is used in the field of solar cells, light-emitting diodes, lasers, and photodetectors. This material has caught the attention of scientists for its potential application as high-energy radiation detectors and scintillators due to their excellent light yield, mobility lifetime product ($\mu\tau$), and X-ray sensitivity [13]. The unique properties of perovskites come from their special structure.

### 3.1 The structure and stability of perovskite

Perovskite is a family of compounds that share the same crystal structure as $CaTiO_3$ and are often represented by the ideal formula $AMX_3$. In the area of photoelectric, the cation A is typically an organic ammonium ion ($CH_3NH_3^+$), a fluor ammonium ion ($CH(NH_2)^{2+}$), or a cesium ion ($Cs^+$), while the halide anion X is represented by $Cl^-$, $Br^-$, $I^-$, and the metal cation B by $Pb^{2+}$, $Ge^{2+}$, $Sn^{2+}$, $Cd^{2+}$, $Mn^{2+}$, and others [14]. These compounds form $BX_6^{4-}$ octahedral structures by coordinating metal cations B and halide anions X, which are interconnected through shared vertices to create a three-dimensional network structure [15]. Halide ions X can also be partially replaced by other halogen atoms, resulting in mixed perovskites like $ABCl_xBr_{3-x}$, $ABBr_xI_{3-x}$, and others. The band structure of perovskite materials varies depending on the chemical composition of $ABX_3$, with the A-site cation being highly ionized and contributing little to the band edge [16]. In contrast, the $BX_6^{4-}$ octahedral structure directly determines the band structure of perovskites [17].

The crystal structure of perovskite is highly flexible, but it needs to meet certain conditions to keep stable. The definitions of Goldschmidt tolerance factor (t) and

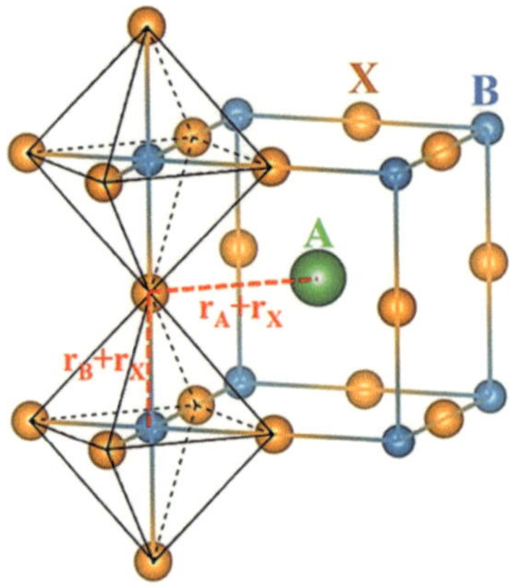


$$t = \frac{r_A + r_X}{\sqrt{2}(r_B + r_X)}$$

$$u = \frac{r_B}{r_X}$$

$$\eta = \frac{V_A + V_B + 3 \cdot V_X}{a^3}$$

**Figure 3.**
*Definitions of the structural factors t, μ, η in a cubic phase perovskite crystal structure [18].*

the octahedral factor (μ) are shown in the **Figure 3**; both are used to describe the stability of the perovskite structure [19]. Among them, t represents the degree of distortion of the perovskite structure and is related to the ionic radii r of A, B, X, satisfying the formula (1–1), μ is directly calculated of the octahedral complex $BX_6^{4-}$ as shown in the formula (1–2). Atomic packing fraction (η) can be described by formula (1–3), where $V_A$, $V_B$, and $V_X$ are atomic volumes of A, B, and X and a is the lattice constant of the cubic cell. Wan-Jian Yin's team studied 138 kinds of perovskite compounds using first-principles calculations. They found a linear correlation between decomposition energies and the descriptor $(\mu + t)^{\eta}$, indicating a notable thermodynamic stability trend that can be used to predict stable perovskite structures [18].

### 3.2 Synthesis method of fully inorganic perovskite quantum dots

The crystal quality of perovskite quantum dots is the most critical factor determining the efficiency of scintillator devices. Optimizing synthesis methods to achieve control over the morphology, size, and structure of nanocrystals is also crucial. Currently, there are several main methods for preparing perovskite quantum dots:

The hot-injection is the most classic synthesis method for quantum dots, which involves rapidly injecting one type of precursor into a high-temperature solvent containing other precursors and necessary ligands. After the injection of the precursor, nucleation and growth occur rapidly. This method can produce high-quality crystals with a narrow size distribution. The hot-injection synthesis method allows for the control of the size, morphology, and structure of the growing nanocrystals, mainly by controlling: (1) the reaction temperature at injection; (2) reaction time; (3) ligands or activators; (4) precursor concentration; (5) the ratio of precursor components. As illustrated in **Figure 4a** [21], the hot-injection method was first used in 2015 by the Protesescu research group to synthesize perovskite $CsPbX_3$ [22]. The primary method involved synthesizing cesium oleate as a precursor, which was then injected into a high-boiling solvent containing $PbX_2$ at temperatures ranging from 140 to 200°C. The final nanocrystal size could be controlled by adjusting the temperature, thus tuning the luminescence peak position based on the quantum size effect. Additionally, by simply changing the proportion of halogens in $PbX_2$, mixed halogen perovskite

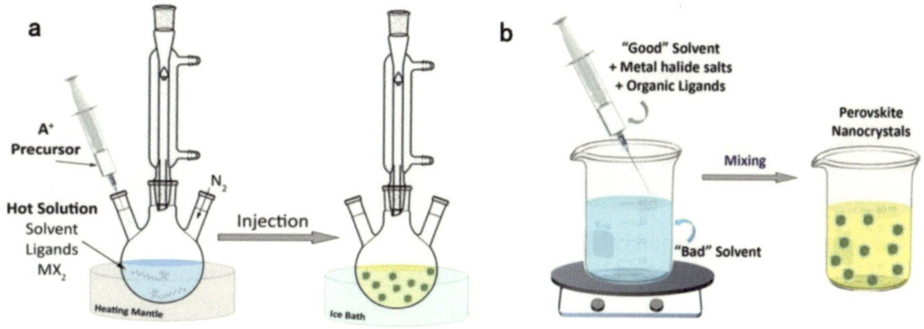


**Figure 4.**
*Diagram of (a) high-temperature hot-injection method; (b) room-temperature recrystallization method [20].*

nanocrystals could be synthesized. Therefore, this method allows for complete and fine-tuned coverage across the entire visible light spectrum.

When ions dissolve in a solvent, their concentration increases with the dissolution of more ions until reaching equilibrium concentration. If the solution becomes supersaturated and non-equilibrium, precipitation and crystallization will spontaneously occur in the solution to restore balance. Moving the system from an equilibrium to a supersaturated non-equilibrium state can be achieved by controlling temperature, evaporating some of the solvent, or mixing in another solvent with lower ion solubility. In chemical synthesis, ligands are often used to further control the crystallization process, allowing for nanoscale control of crystal growth, hence this method is known as ligand-assisted reprecipitation in chemical synthesis. As shown in **Figure 4(b)**, the ligand-assisted reprecipitation method is widely used in the synthesis of perovskites. Typically, precursors (like CsX, $PbX_2$) and ligands are dissolved in a highly polar solvent such as DMSO or DMF, thoroughly mixed and dissolved. Then, a small amount is dropped into a poor solvent for perovskites (like toluene), causing nucleation and growth of perovskites due to the instantaneous shift from a supersaturated non-equilibrium state to equilibrium. This method can be proceeded in the air and is easier for scaleup. However, the size distribution of quantum dots synthesized by this method is larger.

### 3.3 Classification and applications of perovskite scintillator materials

Traditional scintillators have consistently served as the gold standard in performance evaluation for nascent scintillating materials. Herein, they are classified into three cardinal sub-categories: single-crystal, oxide-based, and organic scintillators. Single-crystal scintillators and oxide scintillators normally require high vacuum and high-temperature processing. Organic scintillators offer advantages such as ease of synthesis, low cost, and shorter decay times of approximately 1 ns [23, 24]. However, complexities emerge due to their low density (around 1 g $cm^{-3}$) and low effective atomic number (Z), leading to a drastically reduced detection efficiency for high-energy X-rays [25]. Furthermore, organic dyes exhibit a heightened susceptibility to photobleaching and quenching effects from oxygen, significantly impeding their utilization across various applications.

Over the past few years, perovskite materials, due to their distinctive attributes such as a high photoluminescence quantum yield (PLQY), swift decay time, substantial atomic number, easily tunable luminescence color, and straightforward, cost-effective preparation methods, have been identified as potential scintillator materials [26, 27].

Based on the composition and morphological properties of perovskite, perovskite scintillators can be categorized into the following groups: single-crystal scintillators, two-dimensional (2D) scintillators, and perovskite quantum dots (PQDs) scintillators.

### *3.3.1 Single crystal (SCs) scintillators*

As previously stated, the quality of the scintillating material is a critical determinant in gauging the sensitivity of X-ray detectors. Therefore, excellent scintillators should ideally exhibit characteristics such as significant volume resistivity, diminished dark current, superior crystallinity, and fewer trapping states, which will enhance the $\mu\tau$ value. Notably, an optimal perovskite SC demonstrates small trap states and no grain boundaries, subsequently reducing the likelihood of carrier scattering and thus accelerating the carrier migration rate, ultimately enhancing the overall performance of the device [7].

Additionally, beneficial attributes such as considerable thickness, high density, effective energy response, and a broad absorption cross-section of perovskite SC scintillators have demonstrated enormous potential in X-ray detection. However, the implementation of traditional high-temperature furnace growth techniques for SC scintillators presents challenges including not only high production costs but also inducing doping gradient that leads to uneven light output and a decrease in resolution, precluding large-scale production.

### *3.3.2 Two-dimensional scintillators*

Two-dimensional scintillators represent nanoscale materials, facilitating free-electron movement across only two dimensions. The transition from 3D to 2D structures boosts the quantum confinement effect within perovskite scintillators, producing several unique features [3]:

1. Efficacious scintillation under ambient conditions. Conventionally, substantial 3D perovskites barely facilitate impactful room-temperature scintillation owing to low exciton binding energy engendering prominent thermal quenching. In contrast, for identical materials, 2D structures typically possess higher exciton binding energy than their 3D counterparts, thereby mitigating luminescence efficiency interferences from phonons. Deep exciton energy levels consequently endow them with thermal quenching resistance at room temperature.

2. Swift response and minimal decay. Quantum wells intrinsic to 2D perovskites confine electrons and holes instigated by ionizing radiation, augmenting the overlap of the electron-hole wave function. This sole electronic characteristic amplifies exciton oscillator strength while mitigating 2D perovskites' exciton radiation lifetime.

3. Exceptional environmental stability against ionizing radiation. Layered 2D halide perovskites' stability lies in their unique architecture, featuring inorganic semiconductor layers enveloped by copious insulating organic ones. The existence of potent hydrogen bonds among inorganic and organic units and hydrophobicity characteristic to organic spacers bestows onto scintillators superior resistance against humidity and radiation intensity. Additionally, 2D scintillator synthesis can achieve exceptional thinness to meet X-ray micro-imaging

requirements, assisting in mitigating response time delays and efficiency losses provoked by the self-absorption effect. 2D scintillators are envisaged for future extensive applications [28].

### *3.3.3 Quantum dots scintillators*

Standing in contrast to previously mentioned material structures, PQDs manifest amplified quantum confinement effects, due to unique electronic structures, resulting in superior scintillation performance. Moreover, due to quantum confinement effects, nanocrystal scintillators generate excellent luminescent efficiency. In recent years, Halide PQDs scintillators have been extensively studied for their easy-to-manufacture nature, tunable bandgap within the visible light spectrum, and high X-ray absorption coefficients [1].

The wet chemical solution processibility and controllable thickness of PQDs scintillators make them ideal candidates for flexible or large-area panel detectors within imaging technologies. Furthermore, PQDs solutions can function as inks within manufacturing processes. Regarding X-ray scintillation applications, materials' tolerance to high-power X-ray radiations is fundamental in constructing medical detectors. Various all-inorganic halide perovskites have steered an increasing research interest in scintillation applications, attributed to their swift response and high detection efficacy toward X-rays.

The emerging nanoscale perovskites possess commendable manufacturability when fabricating scintillating thin films, favoring the construction of flexible devices. Furthermore, owing to quantum and small-size effects, perovskite PQDs scintillators showcase a unique set of electromagnetic and extensively extended optical characteristics. It is notable that while material downsizing to the nanoscale can augment

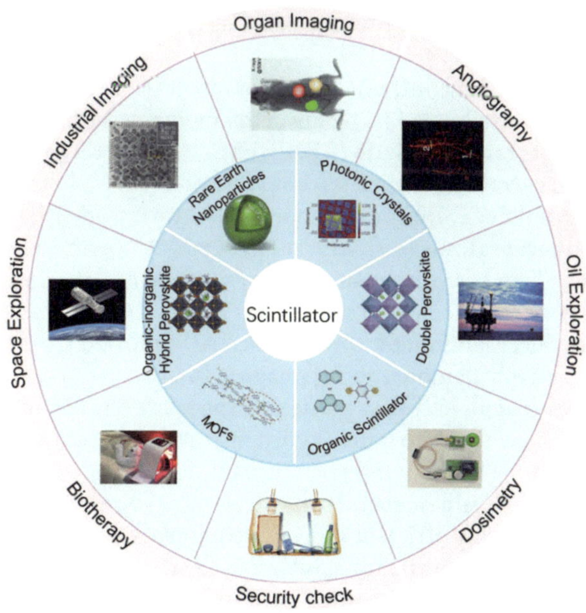


**Figure 5.**
*Scintillator materials and application fields used in X-ray indirect detectors [29].*

scintillation performance immensely, it might concurrently induce a reduction in effective mass density, thereby curtailing the X-ray stopping power. Fortunately, researchers have devised effective solutions to this predicament through novel manufacturing methodologies. For instance, integrating nanocrystals with glassy media has realized high stability and high-resolution X-ray imaging.

Although PQDs scintillators have observed widespread use in detection and imaging, several leaps persist toward achieving practical applications. Numerous technical issues await solutions—structural optimization, detection efficiency improvement, and cost reduction to name a few. Typically, thicker active materials facilitate effective X-ray photon absorption; however, inefficient carrier collection could compromise device sensitivity. Furthermore, despite PQDs scintillators' potential for conjunction with silicon detectors, strong re-absorption phenomena inherent between them result in decreased silicon detector efficiency (**Figure 5**) [30].

## 4. Progress in the development of perovskite quantum dot scintillators

Perovskite single crystals' luminescent yield at ambient conditions is relatively low (less than 1000 photons/MeV) [31, 32], gaining merit solely at lowered temperatures [33]. Research postulates that perovskite single crystals undergo considerable thermal quenching at room temperature, leading to non-radiative recombination of most carriers rendering low luminescent yield [34].

Leveraging quantum confinement effects, Asai's research team [28, 35] fabricated two-dimensional perovskites. This approach enhanced the exciton binding energy of the crystals, subdued thermal quenching, and led to the creation of various "quantum scintillators" exhibiting ultrafast responses. Koshimizu's team further advanced these two-dimensional perovskite scintillators by synthesizing the quintessential $(Phe)_2PbBr_4$ crystals, Phe = $C_6H_5(CH_2)_2NH_3$. They realized sub-nanosecond time resolutions scintillator with decay lifetime of 7.4 ns [36], and a luminescent yield of 14,000 photons/MeV [37].

The team also explored the influence of disparate organic groups and inorganic ions on the scintillator attributes [3]. Noteworthy members of the two-dimensional perovskite scintillators family include $(EDBE)PbCl_4$ [31], $(PEA)_2PbBr_4$ [38], and $(BA)_2PbBr_4$ [39], all exceeding a luminescent yield of 10,000 photons/MeV and exhibiting nanosecond decay lifetimes. Some exploratory studies leveraged the incorporation of fluoride atoms [40], organic groups [41], or alternative metallic ions [42, 43] to enhance these two-dimensional perovskite scintillators with satisfactory progress. Lead-free two-dimensional perovskite scintillators have surfaced as a novel research direction. Successful material synthesis [43] followed the replacement of $Pb^{2+}$ with $Sn^{2+}$, effectively demonstrating imaging [4].

In 2015, Protesescu et al. [22] first pioneered the synthesis of inorganic perovskite nanocrystals ($CsPbX_3$, X refers to halide) with near 90% photoluminescence quantum efficiency (PLQY). This material structure reinforced the quantum confinement effects, displaying high luminescence rates and minimal non-radiative recombination, thus serving as a highly efficient light-emitting material. Research teams recognized the material's potential for scintillation and studied $CsPbX_3$ inorganic perovskite nanocrystals for X-ray imaging applications, resulting in elementary indirect X-ray imagers [11]. Exemplarily, the green-emitting $CsPbBr_3$ demonstrated the best performance, prompting researchers to explore X-ray imaging performances based on this scintillator (**Figure 6**) [44].

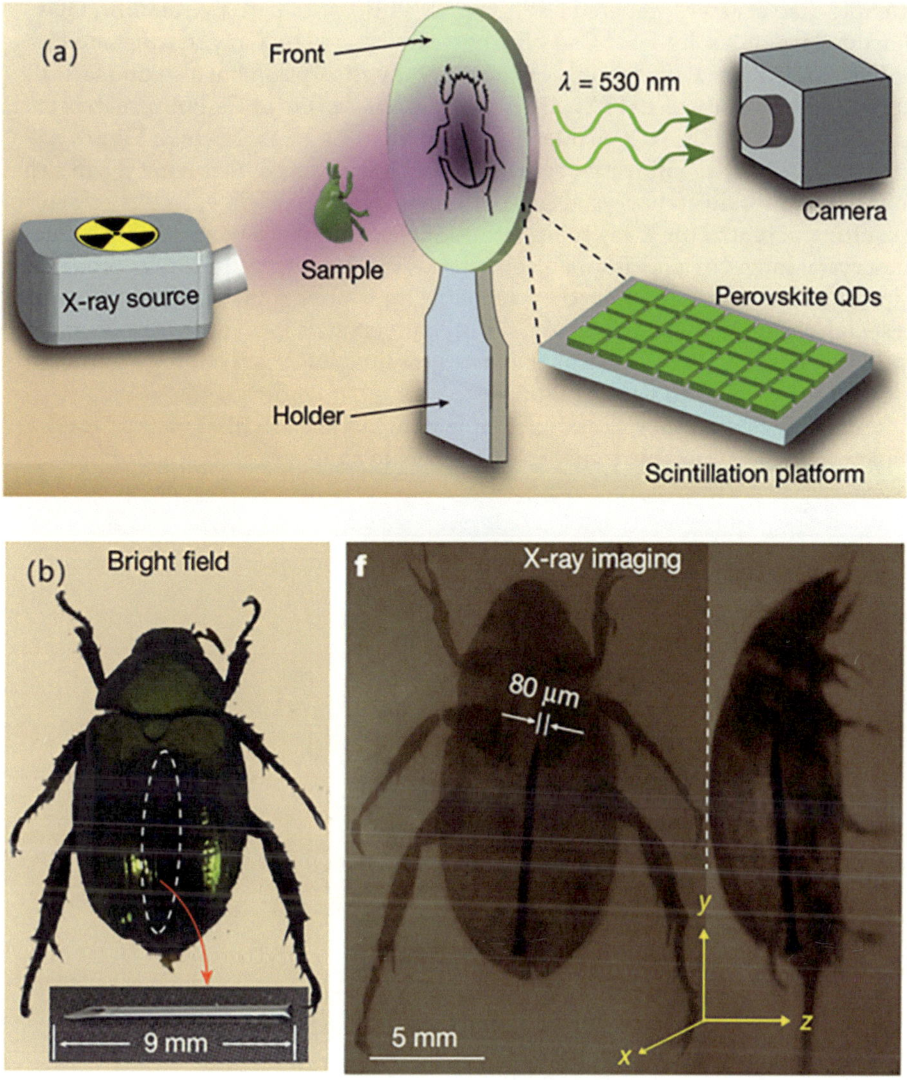


**Figure 6.**
*(a) Ultrasensitive X-ray sensing and radi-ography using $CsPbBr_3$ nanocrystals. (b) Image of the sample under X-ray [6].*

PQDs scintillators exhibit several significant challenges: (1) A pervasive deficiency of methodologies allowing for the large-scale fabrication of scintillator films; (2) compromised stability, which manifests in performance degradation when subjected to heat and humidity in the ambient environment; (3) limited imaging resolution with the inherent opacity of thick films persistently hampering their overall imaging efficacy.

To solve these challenges, researchers have studied various solutions. Wang et al. [45] disclosed a room-temperature process for fabricating high-content $CsPbBr_3$ nanosheet colloidal scintillators by directly mixing varied amount of $CsPbBr_3$ QDs with acrylate-based resin, the resulting 200 um thick $CsPbBr_3$ QD-resin layers were laminated with 100 μm thick barrier films on both sides as a water vapor barrier (as shown in **Figure 7**), varied polymer and resins have been applied as matrix for PQDs follows this advancement. Similarly, based on the "emitter-in-matrix" design principle, Cao et al. [1] prepared $CsPbBr_3$@$Cs_4PbBr_6$ with emissive $CsPbBr_3$ QDs embedded inside a solid-state $Cs_4PbBr_6$ host is subjected to X-ray sensing and imaging. The $Cs_4PbBr_6$ matrix not only enhances X-ray absorption but also dramatically improves the stability of $CsPbBr_3$ QDs. The low absorption of $Cs_4PbBr_6$ matrix to the emission from $CsPbBr_3$ NCs enables efficient light output, as shown in **Figure 8**.

In recent years, an innovative research trend has surfaced, involving the in situ growth of perovskite nanocrystals within vitreous matrices [45, 47], thus presenting an exciting potential for X-ray scintillators. Such an approach considerably bolsters nanocrystal longevity and attains a commendable imaging resolution, approximately 15 lp/mm. Taking this a step further, a notable innovative solution was proposed by a research team [48] who dissolved $CsPbBr_3$ nanocrystals in an organic solvent, the resultant transparent solution was then used for imaging, representing significant strides in the field, as shown in **Figure 9**.

Recently, Wu et al. [3] successfully implemented a suction filter method for the preparation of an ultra-thin transparent perovskite nanocrystal scintillator film.

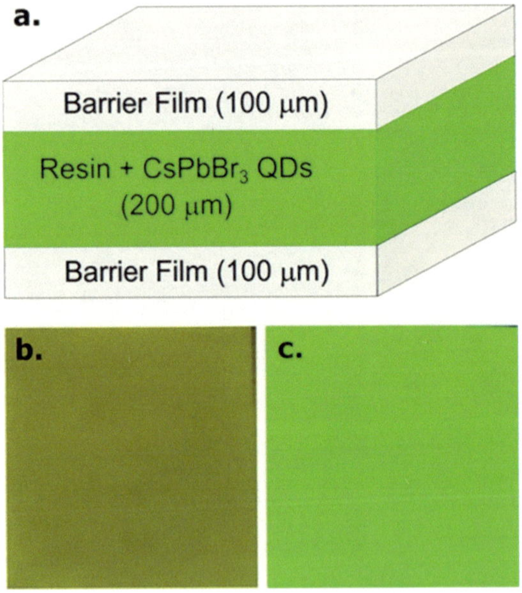


**Figure 7.**
*(a) Schematic representation of the sample structure with the $CsPbBr_3$ QDs mixed in resin between two barrier films. Image of 15% $CsPbBr_3$ QD concentration in resin under (b) visible light and (c) UV-light excitation [46].*

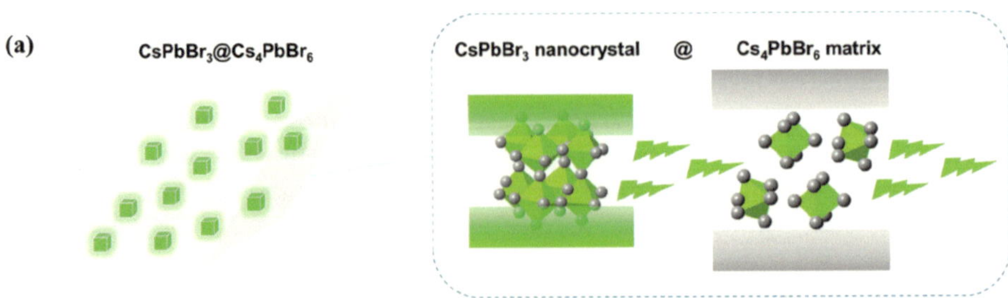


**Figure 8.**
*Schematic view of the band gap of the $CsPbBr_3$ core and the $Cs_4PbBr_6$ matrix [1].*

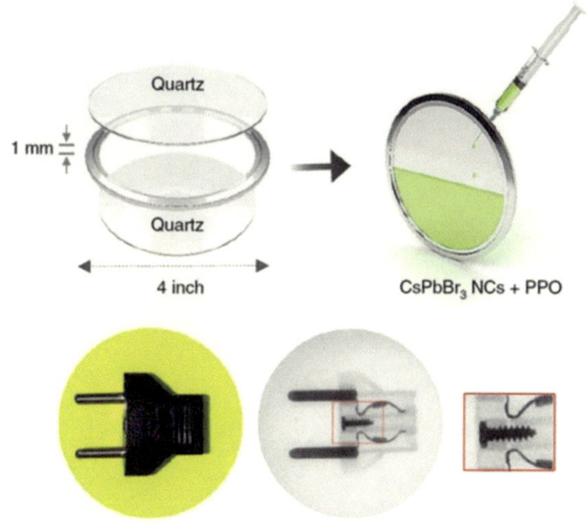


**Figure 9.**
*Schematic diagram of $CsPbBr_3$ liquid scintillators and corresponding imaging demonstration [48].*

Incorporating $Ce^{3+}$ ions as dopants leads to a considerable surge in light yield, enabling a resolution as high as 862 nm for X-ray micro-imaging. With continuous investigations being carried out in the field, some scientists [49] have identified a critical deficit impeding the efficacy of direct band gap perovskite scintillators—self-absorption. Introducing efficient energy transfer in PQDs film could successfully restrain self-absorption to a very low degree. PQDs with varied particle size have varied conduction band (CB) and valence band (VB) positions, an X-ray absorption layere composed of a wider sized distribution of QDs was prepared by Li et al. [26]. The scintillator performance of PQDs mixture with different sizes but at an appropriate ratio was studied. The successful engineering of absorption-emission overlapping of varied size $CsPbBr_3$ QDs avoids the emission loss by reabsorption essentially and ensures the maintenance of high quantum yields of $CsPbBr_3$ QDs. The engineered $CsPbBr_3$ QDs also show more efficient radiative recombination than conventional QDs due to more efficient charge transfer between varied size $CsPbBr_3$ QDs. Instead of using varied size QDs, Gandini et al. [50] fabricated a nanocomposite scintillator by embedding $CsPbBr_3$ QDs together with conjugated organic dyes into the PMMA matrix. As shown in **Figure 10**, the highest occupied molecular orbital (HOMO) and lowest unoccupied molecular orbital (LUMO) levels of the organic dye and $CsPbBr_3$ QDs were perfectly matched, providing an effective energy/charge transfer channel. The emission spectrum of $CsPbBr_3$ QDs is directly overlapped with the main absorption spectrum of the dye, where the emitted photons from $CsPbBr_3$ QDs could further excite the dye molecule, thus reducing the self-absorption of $CsPbBr_3$ QDs. This induced energy transfer between organic molecules and perovskite can effectively increase the utilization efficiency of photons.

Recently, a research team [51] conducted an analysis of $CsPbBr_3$ nanocrystal scintillator films, aiming to determine the optimal thickness to minimize the influence of self-absorption effects. This strategic move found a harmonious balance between X-ray absorption, light yield, and imaging resolution.

Pioneering luminescent principles beyond band gap recombination has motivated significant academic inquiries. A luminary example is a rare-earth ion orbital electron transition-based double perovskite scintillator; $Cs_2NaTbCl_6$ as reported by

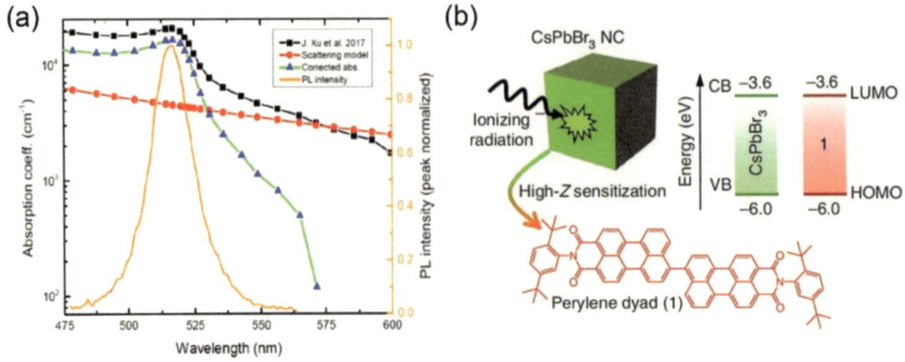


**Figure 10.**
*(a) Self-absorption effect of $CsPbBr_3$ [49]; (b) $CsPbBr_3$ is coupled with organic luminescent materials to form plastic scintillators [50].*

Hu et al. [52]. Despite recording an exceptional light yield of up to 46,600 photons/MeV, the prolonged decay period and severe afterglow may preclude a direct application within real-time X-ray or CT imaging systems.

Furthermore, consequential exploration involves applying diverse element doping to establish multi-modal luminescent scintillators [53] and cerium-doped double perovskite scintillators [54]. Besides this, Zhu and colleagues [55] reported on a lead-free double perovskite scintillator [$Cs_2Ag_{0.6}Na_{0.4}In_{1\text{-}y}Bi_yCl_6$], based on the promising principle of self-trapped exciton (STE) luminescence. Fueling high light yield coupled with a sizeable Stokes shift, the decay duration of this variety of scintillators is markedly shorter, substantially lower than the luminous lifetime of rare-earth ions.

Finally, diminutive halide scintillators leveraging STE luminescence are gradually capturing research attention. They include, prominently, $Rb_2CuBr_3$ [56] and $Cs_3Cu_2I_5$ [57], both of which demonstrate an enormously high light emission power, marking them out as a vanguard family of scintillators.

## 5. Challenges and remedies in perovskite QDs scintillators implementation

Over recent years, extensive research has been carried out on metal halide perovskite scintillators for application in X-ray detection and imagery. Despite this, a multitude of technical hurdles persist, preventing these from practical utilization. These obstacles predominantly encompass reducing self-absorption and amplifying luminescent efficacy, bolstering stability, curtailing production expenses, optimizing production on a larger scale, curbing light scattering, and augmenting imaging resolution [58, 59].

1. Improving light yield: the light yield of the scintillator is inversely proportional to the band gap, so adjusting the bandgap by halogen replacement or B site doping is an efficient way to increase the light yield. Other strategies, including modulation of the exciton confinement effect, minimization of self-absorption in PQDs, constructing energy-transfer channel, and defect modification are also widely studied to improve the light yield.

Many strategies could be adopted to restrain the self-absorption of scintillators. For instance, reducing the scintillator's thickness may considerably mitigate the efficiency losses and response time due to self-absorption. Rare-earth ions (e.g., $Eu^{3+}$, $Yb^{3+}$, $Tb^{3+}$) exhibit intrinsic 4f–4f transitions with large Stokes shifts, introducing these ions into perovskite structures could effectively enlarge the Stokes shift of perovskite scintillators, which leads to a decent scintillation efficiency. Another strategy to lower the self-absorption of the perovskite scintillators is to mix them with molecules with matched energy levels. During the charge transfer of the scintillating processes, the transport charge carriers can be captured by both defects and impurities in the crystal lattices, which can lead to an increase in non-radiative recombination. Therefore, improving crystal quality and reducing defects in the scintillator (such as impurity doping and surface passivation) are also important strategies to enhance the scintillating performance. Furthermore, an appropriate integration of halogen elements could serve to proliferate light transmission and minimize self-absorption effects [60, 61].

2. Elevating stability. The stability quandaries of metal halide perovskite scintillators include both chemical resilience and tolerance to radiation. The primary issue of chemical stability pertains to the sensitivities of perovskites to environmental stimuli such as water, light, and heat. As for radiation tolerance, it refers to the ability of perovskites to endure high-energy radiation. Given the good defect tolerance of perovskites, disruptions in their ionic characterization could easily lead to surface defects. Hence, attempts to enhance perovskite stability may focus on two fronts—augmenting innate structural stability by amending the composition of perovskites and improving crystal development, and deploying chemical strategies to passivate surface anomalies.

3. Mitigating costs and enabling large-scale production. The inherent solubility of metal halide perovskites makes them amenable to mass synthesis. Yet, many issues, such as compatibility, efficiency retention upon large-scale expansion, and alignment with extant processing and manufacturing apparatus, linger before they can be adapted to practical usage. Prior literature suggests that merging perovskite nanocrystals with polymeric materials—using polymers as substrates, coating materials, or as a growth medium—presents as an efficacious strategy to realize large-scale fabrication.

4. Restraining light scattering to augment imaging resolution. A challenge that metal halide perovskites confront pertains to the diminution of imaging resolution as a consequence of light scattering. This phenomenon prompts a photon cross-talk among neighboring luminescent centers, marking a decrement in resolution [62]. Moreover, internal light scattering within the scintillator can result in impaired transparency, hence impacting the imaging resolution. Varied strategy such as waveguide effect and circularly polarized radioluminescence could better confine the propagation directions of emission photons. In addition, resourceful avenues must be explored to suppress light scattering in nanocrystals—for example, by enhancing the crystallinity and dispersion of nanocrystals, obviating nanocrystal congregation, or regulating the size distribution of nanocrystals to facilitate homogeneous crystallization.

Fabricating all-inorganic perovskite quantum dots into composite materials is an effective method to enhance their performance. As early as 2012, Kojima et al.

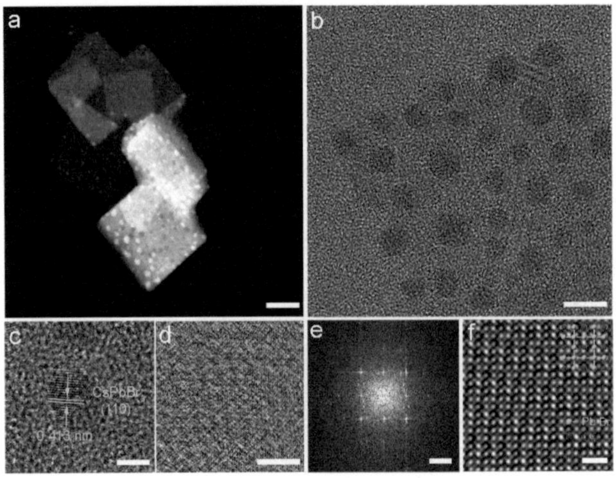


**Figure 11.**
*HAADF-STEM image of (a) $CsPbBr_3$@MOF-PCN-33, (b) TEM image of $CsPbBr_3$@PCN-33(Fe) with the dark dots are $CsPbBr_3$ QDs. Scale bars: a, 25 nm; b, 10 nm [65].*

synthesized $MAPbBr_3$ nanocrystals in situ within mesoporous $Al_2O_3$ or $ZrO_2$ templates, utilizing the mesoporous confinement to limit the growth of the nanocrystals, eventually producing highly luminescent and stable PQDs with a particle size of about 5 nm [63]. It is precisely because of the solubility processing characteristics of PQDs that they can be easily combined with mesoporous inorganic templates. Combining metal halide perovskites with inorganic mesoporous oxides can significantly enhance their stability, but the preparation process is usually complex, and the manufacturing cost is high. Moreover, the optical performance of PQDs is impaired, hindering their further practical application [64].

Besides inorganic encapsulation, organics include polymers and metal-organic frameworks (MOFs). Polymers, with their network structure built on polymer chains, can provide a natural level of nanoscaffold for the growth of nanocrystals, so in situ crystallization processes can be used to prepare PQDs nanocomposite as shown in **Figure 11**. However, when selecting polymers, the following points should be considered: they should have the same solvent solubility as the perovskite precursors to ensure effective and thorough mixing; strong interactions should be established between the polymer and perovskite; the polymer should have compatible dielectric properties with the device to be assembled; and it should have a certain size of pore structure.

## 6. Summary and outlook

This comprehensive review has examined the development and current state of perovskite quantum dot scintillators, speculating also on the potential for future advancements. Traditional scintillators, though foundational to the field, display shortcomings that perovskite counterparts have the potential to amend, including cost-efficiency, moisture resistance, and detection capabilities.

The advent of perovskite single-crystal, two-dimensional, and nanocrystalline scintillators has opened up new avenues in the realm of X-ray detection and imaging, attributable to their excellent photoluminescent quantum yield, efficient luminescence, achievable fabrication, and an adjustable bandgap in the visible light spectrum. Considerations regarding high $u\tau$ value maintenance and structural optimization, however, remain paramount.

Significant progress and breakthroughs have been seen in the development of various types of perovskite quantum dot scintillators, from the pioneering fabrication of two-dimensional perovskites, demonstrating distinct photoluminescent properties, to the potential of lead-free two-dimensional perovskite scintillators, exhibiting promising synthesis capabilities and photoluminescent quantum efficiency.

Despite the significant progress made, there are still numerous technical challenges hindering the practical utilization of metal halide perovskite scintillators. These primarily include achieving high luminescent efficiency while minimizing self-absorption, ensuring ultra-fast response times and attenuation, maintaining stability, managing production costs, adapting for large-scale production, reducing light scattering, and enhancing imaging resolution. Innovative solutions are thus necessitated to address these challenges, with potential leadways including thickness reduction of the scintillator, the introduction of defect passivation and interface modification, halogen elements, structural stability enhancements, and the utilization of polymeric materials for scalable production.

In conclusion, the exploration and optimization of perovskite quantum dot scintillators for X-ray imaging and dose monitoring represent an emerging research field. Flexible scintillator films broaden the application of bendable X-ray imaging, releasing the frequency and related risk for radiation damages, enabling more accurate real-time dose monitoring for radiotherapy and tomography.

## Notes/thanks/other declarations

The authors acknowledge support from the Cities partnerships Programme (Rome seed fund) from UCL Global Engagement Office. The European Commission supported this work under the H2020 HI-ACCURACY project (grant agreement ID: 862410).

## Author details

Daihao Li[1], Tianming Sun[1], Stephen Kearney[2], Robert Moss[2] and Mingqing Wang[1*]

1 Institute for Materials Discovery, University College London, London, UK

2 Department of Medical Physics and Biomedical Engineering, University College London, London, UK

*Address all correspondence to: mingqing.wang@ucl.ac.uk

## References


[1] Cao F, Yu D, Ma W, Xu X, Cai B, Yang YM, et al. Shining emitter in a stable host: Design of halide perovskite scintillators for X-ray imaging from commercial concept. ACS Nano. 2020;**14**(5):5183-5193. DOI: 10.1021/acsnano.9b06114

[2] Lu L, Sun MZ, Lu QY, Wu T, Huang BL. High energy X-ray radiation sensitive scintillating materials for medical imaging, cancer diagnosis and therapy. Nano Energy. 2021;**79**(1):105437. DOI: 10543710.1016/j.nanoen.2020.105437

[3] Wu X, Guo Z, Zhu S, et al. Ultrathin, transparent, and high density perovskite scintillator film for high resolution X-ray microscopic imaging[J]. Advanced Science. 2022;**9**(17):2200831

[4] Cao J, Guo Z, Zhu S, Fu Y, Zhang H, Wang Q, et al. Preparation of lead-free two-dimensional-layered $(C_8H_{17}NH_3)_2SnBr_4$ perovskite scintillators and their application in X-ray imaging. ACS Applied Materials & Interfaces. 2020;**12**(17):19797-19804. DOI: 10.1021/acsami.0c02116

[5] Overdick M, Bäumer C, Engel KJ, Fink J, Hermann C, Krüger H, et al. Status of direct conversion detectors for medical imaging with X-rays. IEEE Transactions on Nuclear Science. 2010;**57**(5):2995-2995. DOI: 10.1109/Tns.2010.2079470

[6] Chen Q, Wu J, Ou X, et al. All-inorganic perovskite nanocrystal scintillators[J]. Nature; 2018;**561**(7721): 88-93

[7] Robbins DJ. On predicting the maximum efficiency of phosphor systems excited by ionizing-radiation. Journal of the Electrochemical Society. 1980;**127**(12):2694-2702. DOI: 10.1149/1.2129574

[8] Lempicki A, Wojtowicz AJ, Berman E. Fundamental limits of scintillator performance. Nuclear Instruments and Methods in Physics Research Section A: Accelerators, Spectrometers, Detectors and Associated Equipment. 1993;**333**(2-3):304-311. DOI: Doi 10.1016/0168-9002(93)91170-R

[9] Dorenbos P. Scintillation mechanisms in $Ce^{3+}$ doped halide scintillators. Physica Status Solidi (a). 2005;**202**(2):195-200. DOI: 10.1002/pssa.200460106

[10] Kumar V, Luo ZP. A review on X-ray excited emission decay dynamics in inorganic scintillator materials. Photonics-Basel. 2021;**8**(3):10.3390/photonics8030071

[11] Chen QS, Wu J, Ou XY, Huang BL, Almutlaq J, Zhumekenov AA, et al. All-inorganic perovskite nanocrystal scintillators. Nature. 2018;**561**(7721):88-93. DOI: 10.1038/s41586-018-0451-1

[12] Sysoeva E, Tarasov V, Zelenskaya O. Comparison of the methods for determination of scintillation light yield. Nuclear Instruments and Methods in Physics Research Section A: Accelerators, Spectrometers, Detectors and Associated Equipment. 2002;**486**(1-2):67-73. DOI: 10.1016/S0168-9002(02)00676-9

[13] Jana A, Cho S, Patil SA, Meena A, Jo Y, Sree VG, et al. Perovskite: Scintillators, direct detectors, and X-ray imagers. Materials Today. 2022;**55**:110-136. DOI: 10.1016/j.mattod.2022.04.009

[14] Oku T. Crystal structures of perovskite halide compounds used

for solar cells. Reviews on Advanced Materials Science. 2020;**59**(1):264-305. DOI: 10.1515/rams-2020-0015

[15] Zhang Z, Zhao X, Wang T, Zhao Y, Shen C, Trevor M, et al. Research progress of solar cells based on organic-inorganic hybrid perovskites methylamine lead halide. Energy and Environment Focus. 2014;**3**(4):354-359. DOI: 10.1166/eef.2014.1122

[16] Lee MM, Teuscher J, Miyasaka T, Murakami TN, Snaith HJ. Efficient hybrid solar cells based on meso-superstructured organometal halide perovskites. Science. 2012;**338**(6107):643-647. DOI: 10.1126/science.1228604

[17] Xiao ZW, Zhou YY, Hosono H, Kamiya T, Padture NP. Bandgap optimization of perovskite semiconductors for photovoltaic applications. Chemistry – A European Journal. 2018;**24**(10):2305-2316. DOI: 10.1002/chem.201705031

[18] Sun Q, Yin W-J. Thermodynamic stability trend of cubic perovskites. Journal of the American Chemical Society. 2017;**139**(42):14905-14908. DOI: 10.1021/jacs.7b09379

[19] Subhani WS, Wang K, Du MY, Liu SF. Goldschmidt-rule-deviated perovskite $CsPbIBr_2$ by barium substitution for efficient solar cells. Nano Energy. 2019;**61**:165-172. DOI: 10.1016/j.nanoen.2019.04.066

[20] Shamsi J, Urban AS, Imran M, De Trizio L, Manna L. Metal halide perovskite nanocrystals: Synthesis, post-synthesis modifications, and their optical properties. Chemical Reviews. 2019;**119**(5):3296-3348. DOI: 10.1021/acs.chemrev.8b00644

[21] Pan G, Bai X, Yang D, Chen X, Jing P, Qu S, et al. Doping lanthanide into perovskite nanocrystals: Highly improved and expanded optical properties. Nano Letters. 2017;**17**(12):8005-8011. DOI: 10.1021/acs.nanolett.7b04575

[22] Protesescu L, Yakunin S, Bodnarchuk MI, Krieg F, Caputo R, Hendon CH, et al. Nanocrystals of cesium Lead halide perovskites ($CsPbX_3$, X = Cl, Br, and I): Novel optoelectronic materials showing bright emission with wide color gamut. Nano Letters. 2015;**15**(6):3692-3696. DOI: 10.1021/nl5048779

[23] Galunov NZ, Khromiuk IF, Tarasenko OA. Features of pulse shape discrimination capability of organic heterogeneous scintillators[J]. Nuclear Instruments & Methods in Physics Research. 2020;**949**(Jan.1):162870.1-162870.7. DOI: 10.1016/j.nima.2019.162870

[24] Kagami K, Koshimizu M, Fujimoto Y, Kishimoto S, Haruki R, Nishikido F, et al. High-energy X-ray detection capabilities of Hf-loaded plastic scintillators synthesized by sol-gel method. Journal of Materials Science-Materials in Electronics. 2020;**31**(2):896-902. DOI: 10.1007/s10854-019-02597-y

[25] Gandini M, Villa I, Beretta M, Gotti C, Imran M, Carulli F, et al. Efficient, fast and reabsorption-free perovskite nanocrystal-based sensitized plastic scintillators. Nature Nanotechnology. 2020;**15**(6):462-468. DOI: 10.1038/s41565-020-0683-8

[26] Li XM, Meng CF, Huang B, Yang DD, Xu XB, Zeng HB. All-perovskite integrated X-ray detector with ultrahigh sensitivity. Advanced Optical Materials. 2020;**8**(12):2000273. DOI: 10.1002/adom.202000273

[27] Li Z, Jin Z. HI hydrolysis-derived intermediate as booster for $CsPbI_3$ perovskite: From crystal structure, film

fabrication to device performance. Journal of Semiconductors. 2020;**41**(5):051202. DOI: 10.1088/1674-4926/41/5/051202

[28] Shibuya K, Koshimizu M, Murakami H, Muroya Y, Katsumura Y, Asai K. Development of ultra-fast semiconducting scintillators using quantum confinement effect. Japanese Journal of Applied Physics Part 2-Letters & Express Letters. 2004;**43**(10B):L1333-L1336. DOI: 10.1143/JJAP.43.L1333

[29] Yang H, Huang W, Liu X. All-inorganic perovskite nanocrystal scintillator[J]. SCIENCE FOCUS. 2022;**17**(6):20. DOI: 10.15978/j.cnki.1673-5668.202206003

[30] Duan J, Wei J, Tang Q, Li Q. Unveiling the interfacial charge extraction kinetics in inorganic perovskite solar cells with formamidinium lead halide ($FAPbX_3$) nanocrystals. Solar Energy. 2020;**195**:644-650. DOI: 10.1016/j.solener.2019.12.001

[31] Birowosuto MD, Cortecchia D, Drozdowski W, et al. X-ray scintillation in Lead halide perovskite crystals. Scientific Reports. 2016;**6**:37254. DOI: 10.1038/srep37254

[32] Kobayashi M, Omata K, Sugimoto S, Tamagawa Y, Kuroiwa T, Asada H, et al. Scintillation characteristics of $CsPbCl_3$ single crystals. Nuclear Instruments and Methods in Physics Research Section A: Accelerators, Spectrometers, Detectors and Associated Equipment. 2008;**592**(3):369-373. DOI: 10.1016/j.nima.2008.04.079

[33] Mykhaylyk VB, Kraus H, Kapustianyk V, Kim HJ, Mercere P, Rudko M, et al. Bright and fast scintillations of an inorganic halide perovskite $CsPbBr_3$ crystal at cryogenic temperatures. Scientific Reports. 2020;**10**(1):10.1038/s41598-020-65672-z

[34] Xie AZ, Nguyen TH, Hettiarachchi C, Witkowski ME, Drozdowski W, Birowosuto MD, et al. Thermal quenching and dose studies of X-ray luminescence in single crystals of halide perovskites. Journal of Physical Chemistry C. 2018;**122**(28):16265-16273. DOI: 10.1021/acs.jpcc.8b03622

[35] Shibuya K, Koshimizu M, Takeoka Y, Asai K. Scintillation properties of $(C_6H_{13}NH_3)_2PbI_4$: Exciton luminescence of an organic/inorganic multiple quantum well structure compound induced by 2.0 MeV protons. Nuclear Instruments and Methods in Physics Research Section B: Beam Interactions with Materials and Atoms. 2002;**194**(2):207-212. DOI: 10.1016/S0168-583X(02)00671-7

[36] Kishimoto S, Shibuya K, Nishikido F, Koshimizu M, Haruki R, Yoda Y. Subnanosecond time-resolved x-ray measurements using an organic-inorganic perovskite scintillator. Applied Physics Letters. 2008;**93**(26):10.1063/1.3059562

[37] Kawano N, Koshimizu M, Okada G, Fujimoto Y, Kawaguchi N, Yanagida T, et al. Scintillating organic-inorganic layered perovskite-type compounds and the gamma-ray detection capabilities. Scientific Reports. 2017;**7**(1):14754. DOI: 10.1038/s41598-017-15268-x

[38] Xie AZ, Maddalena F, Witkowski ME, Makowski M, Mahler B, Drozdowski W, et al. Library of two-dimensional hybrid Lead halide perovskite scintillator crystals. Chemistry of Materials. 2020;**32**(19):8530-8539. DOI: 10.1021/acs.chemmater.0c02789

[39] Li Y, Chen L, Liu B, Jin P, Gao RL, Zhou LD, et al. Scintillation

performance of two-dimensional perovskite $(BA)_2PbBr_4$ microcrystals. Journal of Materials Chemistry C. 2021;**9**(47):17124-17128. DOI: 10.1039/d1tc04072c

[40] Kawano N, Akatsuka M, Nakauchi D, Kimura H, Yanagida T. Scintillation properties of organic-inorganic layered perovskite-type compounds with a methylphenethylamine. Japanese Journal of Applied Physics. 2022;**61**(SB). DOI: 10.35848/1347-4065/ac2038

[41] Kawano N, Nakauchi D, Akatsuka M, et al. Photoluminescence and scintillation characteristics of organic-inorganic layered perovskite-type compounds with a methoxyphenethylamine[J]. Journal of Luminescence. 2022;**241**:118467. DOI: 10.1016/j.jlumin.2021.118467

[42] Onoda D, Akatsuka M, Kawano N, Kato T, Nakauchi D, Kawaguchi N, et al. Effect of Cu doping on photoluminescence and scintillation properties of $(C_6H_5C_2H_4NH_3)_2PbBr_4$. Sensors and Materials. 2022;**34**(2):585-593. DOI: 10.18494/SAM3679

[43] Maddalena F, Xie AZ, Arramel WME, Makowski M, Mahler B, et al. Effect of commensurate lithium doping on the scintillation of two-dimensional perovskite crystals. Journal of Materials Chemistry C. 2021;**9**(7):2504-2512. DOI: 10.1039/d0tc05647b

[44] Hyuck HJ, Hee SD, Kyoung PJ, et al. High-Performance Next-Generation Perovskite Nanocrystal Scintillator for Nondestructive X-Ray Imaging[J]. Advanced Materials. 2018;**30**(40):1801743. DOI: 10.1002/adma.201801743

[45] Wang CY, Lin H, Zhang ZJ, Qiu ZH, Yang HY, Cheng Y, et al. X-ray excited $CsPb(Cl,Br)_3$ perovskite quantum dots-glass composite with long-lifetime. Journal of the European Ceramic Society. 2020;**40**(5):2234-2238. DOI: 10.1016/j.jeurceramsoc.2020.01.016

[46] Maddalena F, Witkowski ME, Makowski M, Bachiri A, Mahler B, Wong Y-C, et al. Stable and bright commercial CsPbBr3 quantum dot-resin layers for apparent X-ray imaging screen. ACS Applied Materials & Interfaces. 2021;**13**(49):59450-59459. DOI: 10.1021/acsami.1c16171

[47] Zhang H, Yang Z, Zhou M, Zhao L, Jiang T, Yang H, et al. Reproducible X-ray imaging with a perovskite nanocrystal scintillator embedded in a transparent amorphous network structure. Advanced Materials. 2021;**33**(40):e2102529. DOI: 10.1002/adma.202102529

[48] Cho S, Kim S, Kim J, Jo Y, Ryu I, Hong S, et al. Hybridisation of perovskite nanocrystals with organic molecules for highly efficient liquid scintillators. Light-Science & Applications. 2020;**9**(1):10.1038/s41377-020-00391-8

[49] Williams RT, Wolszczak WW, Yan XH, Carroll DL. Perovskite quantum-dot-in-host for detection of ionizing radiation. ACS Nano. 2020;**14**(5):5161-5169. DOI: 10.1021/acsnano.0c02529

[50] Gandini M, Villa I, Beretta M, Gotti C, Imran M, Carulli F, et al. Efficient, fast and reabsorption-free perovskite nanocrystal-based sensitized plastic scintillators. Nature Nanotechnology. 2020;**15**(6):462, 10.1038/s41565-020-0683-8

[51] Wang ZF, Sun RJ, Liu NQ, Fan HL, Hu X, Shen DP, et al. X-ray imager of 26-μm resolution achieved by perovskite assembly. Nano Research. 2022;**15**(3):2399-2404. DOI: 10.1007/s12274-021-3808-y

[52] Hu QS, Deng ZZ, Hu MC, Zhao AJ, Zhang YQ, Tan ZF, et al. X-ray scintillation in lead-free double perovskite crystals. Science China-Chemistry. 2018;**61**(12):1581-1586. DOI: 10.1007/s11426-018-9308-2

[53] Zeng ZC, Huang BL, Wang X, Lu L, Lu QY, Sun MZ, et al. Multimodal luminescent $Yb^{3+}$/$Er^{3+}$/$Bi^{3+}$-doped perovskite single crystals for X-ray detection and anti-counterfeiting. Advanced Materials. 2020;**32**(43):10.1002/adma.202004506

[54] Wang Z, Xu X, Wang S, Xu H, Xu W, Zeng Q, et al. Cerium doping double perovskite scintillator for sensitive X-ray detection and imaging. Chemistry. 2021;**27**(35):9071-9076. DOI: 10.1002/chem.202100449

[55] Klein CA. Bandgap dependence and related features of radiation ionization energies in semiconductors. Journal of Applied Physics. 1968;**39**(4):2029, 10.1063/1.1656484

[56] Yang B, Yin L, Niu G, et al. Lead-Free Halide $Rb_2CuBr_3$ as Sensitive X-Ray Scintillator[J]. Advanced Materials. 2019;**31**(44):1904711.1-1904711.8. DOI: 10.1002/adma.201904711

[57] Lian L, Zheng M, Zhang W, et al. Efficient and reabsorption-free radioluminescence in $Cs_3Cu_2I_5$ nanocrystals with self-trapped excitons. Advanced Science. 2020;**7**(11):2000195. DOI: 10.1002/advs.202000195

[58] Pan L, Shrestha S, Taylor N, Nie W, Cao LR. Determination of X-ray detection limit and applications in perovskite X-ray detectors. Nature Communications. 2021;**12**(1):5258. DOI: 10.1038/s41467-021-25648-7

[59] Li ZZ, Zhou FG, Yao HH, Ci ZP, Yang Z, Jin ZW. Halide perovskites for high-performance X-ray detector. Materials Today. 2021;**48**:155-175. DOI: 10.1016/j.mattod.2021.01.028

[60] Yanagida T. Study of rare-earth-doped scintillators. Optical Materials. 2013;**35**(11):1987-1992. DOI: 10.1016/j.optmat.2012.11.002

[61] Xie A, Hettiarachchi C, Maddalena F, Witkowski ME, Makowski M, Drozdowski W, et al. Lithium-doped two-dimensional perovskite scintillator for wide-range radiation detection. Communications Materials. 2020;**1**(1):10.1038/s43246-020-0038-x

[62] Gola A, Ferri A, Tarolli A, Zorzi N, Piemonte C. SiPM optical crosstalk amplification due to scintillator crystal: Effects on timing performance. Physics in Medicine and Biology. 2014;**59**(13):3615-3635. DOI: 10.1088/0031-9155/59/13/3615

[63] Kojima A, Ikegami M, Teshima K, Miyasaka T. Highly luminescent Lead bromide perovskite nanoparticles synthesized with porous alumina media. Chemistry Letters. 2012;**41**(4):397-399. DOI: 10.1246/cl.2012.397

[64] Ahmed GH, Yin J, Bakr OM, Mohammed OF. Successes and challenges of core/shell Lead halide perovskite nanocrystals. ACS Energy Letters. 2021;**6**(4):1340-1357. DOI: 10.1021/acsenergylett.1c00076

[65] Qiao G-Y, Guan D, Yuan S, Rao H, Chen X, Wang J-A, et al. Perovskite quantum dots encapsulated in a mesoporous metal–organic framework as synergistic photocathode materials. Journal of the American Chemical Society. 2021;**143**(35):14253-14260. DOI: 10.1021/jacs.1c05907

Chapter 4

# Sol-Gel Synthesis of $Gd_2O_3:Nd^{3+}$ Nanopowders and the Study of Their Luminescent Properties

*Amir Moussaoui, Dmitry V. Bulyga, Natalia K. Kuzmenko, Alexander I. Ignatev, Sergey K. Evstropiev and Nikolay V. Nikonorov*

## Abstract

The synthesis of gadolinium oxide nanopowders doped with neodymium ($Gd_2O_3:Nd^{3+}$) 30 nm in size was carried out using the citrate sol-gel method that included the simultaneous stabilization of nanoparticles using citric acid and polyvinylpyrrolidone (PVP). This study proposes and assesses a sol-gel synthesis process that involves the use of two different organic stabilizers with different thermal stabilities. Citric acid and polyvinylpyrrolidone were used as organic modifying components, playing a double role in the synthesis process, that is, acting as stabilizers of forming nanoparticles in colloidal solutions and serving as fuel additives in the process of heat treatment of materials. The structural and optical properties of $Gd_2O_3:Nd^{3+}$ were investigated using photoluminescence, FTIR spectroscopy, DTA/TG, scanning electron microscope (SEM), and XRD analysis.

**Keywords:** sol-gel synthesis, absorption spectrum, luminescence, nanoparticle, $Gd_2O_3:Nd^{3+}$

## 1. Introduction

Nanocrystalline materials based on gadolinium oxide ($Gd_2O_3$) are characterized by high luminescent properties and thermal and chemical stability and are promising for various optical, environmental, and medical applications [1].

Liquid methods are often used to synthesize $Gd_2O_3$ nanocrystals: sol-gel process [2]; polymer-salt method [3]; precipitation [4]; hydrothermal method [5]; spray pyrolysis [6], etc. The synthesis process has a substantial influence on the crystal structure and luminescent characteristics of gadolinium oxide nanoparticles. Wet chemical processes are straightforward, efficient, and often used in the production of oxide ceramic materials [7]. Using wet chemical techniques enables a substantial reduction in the temperature required for the synthesis of oxide nanoparticles, consequently accelerating the technical process. $Gd_2O_3$ nanocrystalline powders are synthesized using wet chemical techniques, such as sol-gel or modified Pechini method, and have a stable cubic crystal structure, which is the preferred crystalline form of $Gd_2O_3$ [8].

The sol-gel process, which is widely known and used to obtain various materials, provides their high homogeneity and relatively low synthesis temperatures [9–12]. Thus, in [13], the formation of $Gd_2O_3$ crystals was observed when gels were heat-treated to a temperature of only 400°C, which is significantly lower than the temperatures of technological processes traditionally used in the production of oxide optical materials.

The citrate sol-gel method based on the introduction of citric acid into the initial solutions followed by their heating and formation of homogeneous gels is used to create luminescent nanomaterials based on gadolinium oxide [14].

Citric acid and polyvinylpyrrolidone (PVP) play a dual role in the synthesis process, acting as stabilizers of the formed nanoparticles in colloidal solutions and acting as fuel additive in the process of heat treatment of materials [15, 16].

Citric acid forms chelate compounds with metal ions in solutions and is used to form oxide nanophosphors [15, 17]. During sol-gel synthesis, citric acid molecules undergo evolution at the stage of heating the initial solution and gel formation and completely decompose at $T > 175°C$ [18], that is, temperatures significantly lower than the decomposition temperatures of metal nitrates [19]. For this reason, citric acid is an effective but relatively low-temperature stabilizer of oxide nanoparticles.

PVP is a soluble organic polymer used to stabilize various nanoparticles [20–22]. Thermal decomposition and oxidation of PVP occur at temperatures of 300–550°C, close to the decomposition temperatures of metal nitrates and the formation of oxide nanoparticles. The presence of PVP in the initial solutions has a significant effect on the size and properties of the formed nanoparticles [8, 22]. Compared to citric acid, PVP is a stabilizer that acts at higher temperatures and has a direct role in the formation of oxide nanoparticles.

The aim of the present work is to develop a low-temperature citrate sol-gel synthesis of $Gd_2O_3:Nd^{3+}$ nanopowders with simultaneous application of two organic stabilizers (presented in the table), to study the evolution of nanoparticle structure during their formation and to investigate the luminescent properties of the obtained nanopowders.

## 2. Materials and methods

Aqueous solutions of gadolinium (Gd) and neodymium (Nd) nitrates, citric acid, and PVP were used as starting materials. Aqueous solution of the components was obtained by dissolving powdered reagents in distilled water under vigorous stirring. The mixture of specified volumes of component solution was performed at room temperature. The chemical composition of the obtained liquid mixtures is given in **Table 1**. The mixed solutions were heat-treated at 70°C with stirring.

The processes of material evolution during sol-gel synthesis in the present work were studied via infrared (IR) spectroscopy, differential-thermal and thermogravimetric (DTA-TG), scanning electron microscope (SEM), and XRD analysis.

IR spectra were captured using FTIR spectrometer Bruker ALPHA; DTA-TG analysis was performed using STA 449F1 Jupiter (Netzsch). The MIRA3 TESCAN scanning electron microscope was used to investigate the morphology of the acquired oxide particles. XRD analysis was carried out using diffractometer Rigaku Ultima IV. The average crystallite size was estimated using the Scherrer equation.

| Sample number | Concentration, wt.% | | | | | | |
|---|---|---|---|---|---|---|---|
| | Solution | | | | | Powder | |
| | Water | $Gd(NO_3)_3$ | $Nd(NO_3)_3$ | Citric acid | PVP | $Gd_2O_3$ | Nd |
| 1 | 87.74 | 4.15 | 0.05 | 2.80 | 5.26 | 99 | 1 |

**Table 1.**
*Chemical composition of materials.*

The photoluminescence spectra of obtained powders in the wavelength range of 250–800 nm were investigated using a Perkin–Elmer LS 50B luminescence spectrometer.

The study of emission properties of materials was carried out on an experimental setup including a laser source on a YAG:Nd crystal (wavelength $\lambda$ = 532 nm) generating pulses with duration $\tau$ = 10 ns and energy E = 30 mJ. Acton-300 monochromator (Acton Research) and InGaAs-photodetector ID-44 (Acton Research) were used to record emission spectra. During the research process, $Gd_2O_3$:$Nd^{3+}$ powders were tightly fixed in the space between two plane-parallel polished quartz glass plates, with the thickness of the nanopowder layer between the plates being 150 μm. Laser radiation was focused into a spot with a diameter of about 130 μm on the surface of the plates.

## 3. Results and discussion

### 3.1 FTIR spectra

**Figure 1** displays the infrared absorption spectra of the wet gel 1. The wide and strong absorption band within the 3300–3500 $cm^{-1}$ range is linked to the stretching vibrations of O–H groups. The strong absorption band at 1631 $cm^{-1}$ in the wet gel is attributed to the presence of citrate anions and PVP molecules, which cause vibrations in the carbonyl group C––O. Prior to the heat treatment, the gel contains nitrate anions, which is why the absorption band at 1360 $cm^{-1}$ is attributed to $NO_3$ anions. Additionally, it is important to mention that the wet gel has a weak peak at 537 $cm^{-1}$, which could be attributable to the vibrations of the Gd-O bond.

Previous investigations have shown comparable results in the production of $Y_2O_3$:Eu materials with a similar structure via the citrate sol-gel method [23]. The spectra of cubic $Y_2O_3$ crystals exhibited an absorption band at a wave number of 560 $cm^{-1}$, corresponding to the vibrations of the Y-O bond. This absorption band was identified after drying the gels at 100°C, and its intensity increased with increasing temperatures throughout the heat treatment of the materials [23].

### 3.2 DTA/TG analysis

Heating of the gel during the heat treatment process leads to the decomposition of citric acid, metal salts, and PVP, as well as the formation and growth of oxide crystals. **Figure 2** displays the results of the DTA-TG analysis, which reveals the various processes that take place during the heat treatment of the gel. **Figure 2a** shows that, during the heating of gel, there is a stepwise decrease in the sample mass accompanied by several exothermic effects (**Figure 2b**).

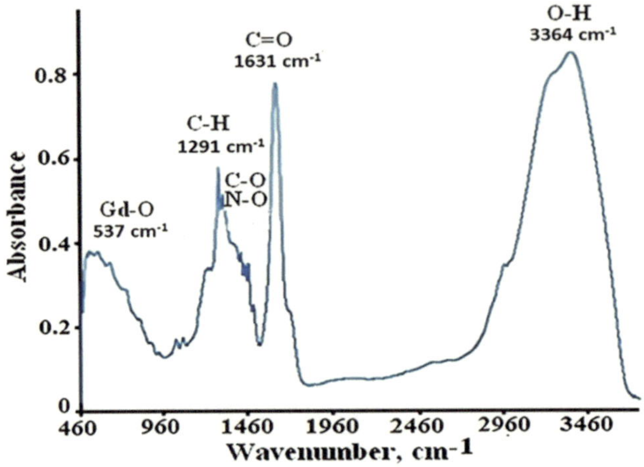


**Figure 1.**
*Absorption spectrum of composite gel in the IR spectral region.*

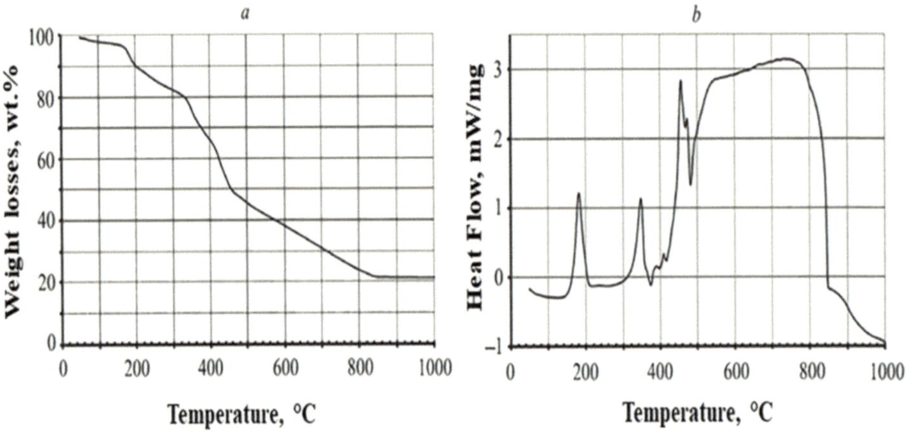


**Figure 2.**
*Results of the thermal gravimetric (a) and differential thermal (b) analyses of the processes occurring during heat treatment of the gel.*

The removal of residual water from the material determines the loss of sample mass and small heat absorption at the initial stages of heat treatment (20–150°C). Citric acid decomposes at a temperature of 175°C, causing a noticeable decrease in the sample mass and the release of heat.

The most significant variations in sample mass changes and exothermic effects are observed in a wide temperature range of 340–850°C. At temperatures of 340–550°C, decomposition of metal nitrates and PVP occurs, accompanied by the significant release of heat and gaseous products.

### 3.3 XRD analysis

**Figure 1** displays the X-ray diffraction (XRD) patterns of $Gd_2O_3$:$Nd^{3+}$ nanocrystalline powders. The powder only comprises peaks that are indicative of the cubic $Gd_2O_3$

structure. Both the amorphous and parasitic crystalline phases, as well as the halo, were absent in the XRD patterns.

Referring to the results presented in Ref. [8], the addition of PVP in the initial solution significantly accelerates the formation of the crystalline phase (**Figure 3**).

The average size of the $Gd_2O_3$ nanocrystals was determined using the Sherrer equation [24]: $d = k\lambda/(\beta \cos \theta)$, where k is the Sherrer constant ($k = 0.9$), $\lambda$ is the x-ray wavelength ($\lambda = 0.15406$ nm) for Cu K$\alpha$, $\beta$ is the full width at half maximum of the corresponding XRD peak, and $\theta$ is the Bragg angle. The average crystallite size of $Gd_2O_3$ was determined to be around 30 nm. According to the findings shown in Ref. [8], the addition of PVP has slight impact on the overall decrease in the average size of the nanocrystals.

### 3.4 SEM analysis

**Figure 4** displays SEM images of $Gd_2O_3$:$Nd^{3+}$ nanocrystalline powder at resolution 20 μm (**Figure 4a**) and 5 μm (**Figure 4b**). The powder is composed of highly aggregated nano-sized $Gd_2O_3$:$Nd^{3+}$ crystals of around 30 nm in diameter. Based on the scanning electron microscope (SEM) image, the average diameter of $Gd_2O_3$ nanocrystals is consistent with the value obtained by calculating it using the Scherrer equation.

The higher magnification image (**Figure 4b**) reveals that the powder is more widely dispersed and comprises small particles, including thin micro-plates. The inclusion of PVP in the sol-gel process results in a material with a highly dispersive form.

### 3.5 Luminescence in UV spectral range

The luminescent properties of the synthesized powders in the UV spectral range are mainly determined by electronic transitions occurring inside the crystalline matrix of $Gd_2O_3$. These transitions are comprehensively explained in [25]. The UV luminescence excitation spectra (**Figure 5a**) exhibit bands with peak wavelengths of 238, 257, 268, and 282 nm. The investigation reported in [26] examined comparable luminescence excitation spectra in $Gd_2O_3$:$Eu^{3+}$ nanophosphors produced by thermochemical and hydrothermal synthesis methods.

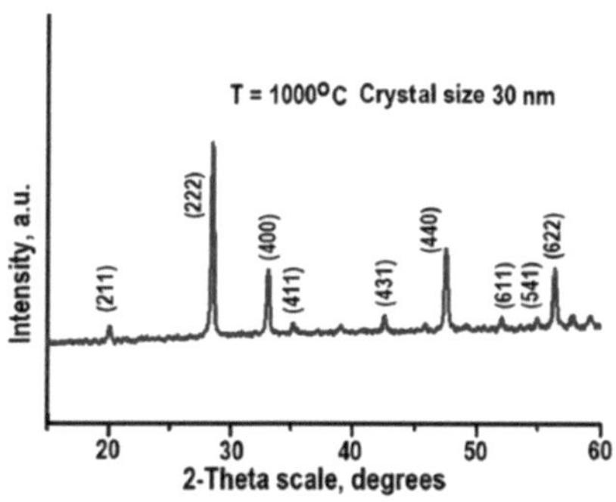


**Figure 3.**
*XRD patterns of the $Gd_2O_3$:$Nd^{3+}$ nanocrystalline powders.*

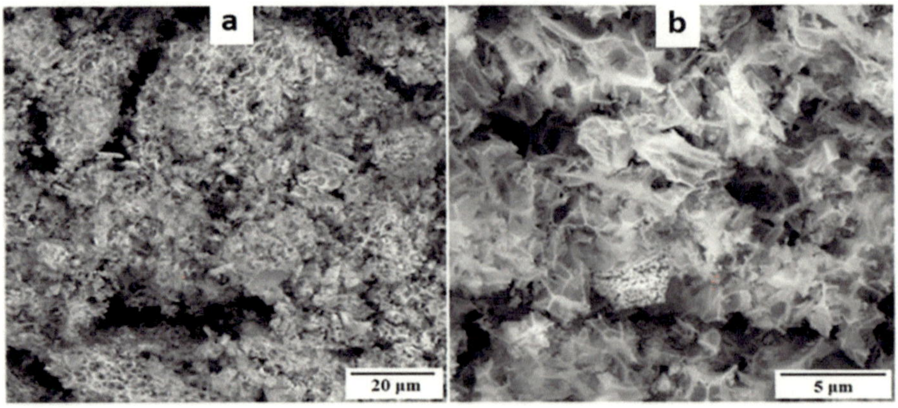


**Figure 4.**
*SEM images of $Gd_2O_3$:$Nd^{3+}$ nanocrystalline powder at resolution 20 μm (a) and 5 μm (b).*

The strongest band with a maximum of $\lambda_{max}$ = 238 nm is linked to the $^6P_{7/2} \rightarrow {}^8S_{7/2}$ transition of $Gd^{3+}$ ions inside the crystalline matrix of $Gd_2O_3$, according to [25]. The band with a maximum of $\lambda_{max}$ = 268 nm corresponds to the electronic transition $^8S_{7/2} \rightarrow {}^6I_{7/2–17/2}$ of $Gd^{3+}$ ions in the crystalline matrix of $Gd_2O_3$ [27]. The charge transfer band with a maximum $\lambda_{max}$ = 257 nm between oxygen ions and $Eu^{3+}$ ions is given in [27]. Upon excitation with a wavelength of λexc = 238 nm, the synthesized powders exhibit luminescence in the ultraviolet region of the spectrum. This luminescence is caused by electronic transitions occurring inside the crystalline matrix of $Gd_2O_3$, as seen in **Figure 5b**.

The experimental findings obtained in this study demonstrate that the use of two stabilizers in the sol-gel process has no impact on the ratio of band intensities exhibited in the luminescence excitation spectra and practically has no effect on the shape of the photoluminescence spectrum in the UV region of the spectrum.

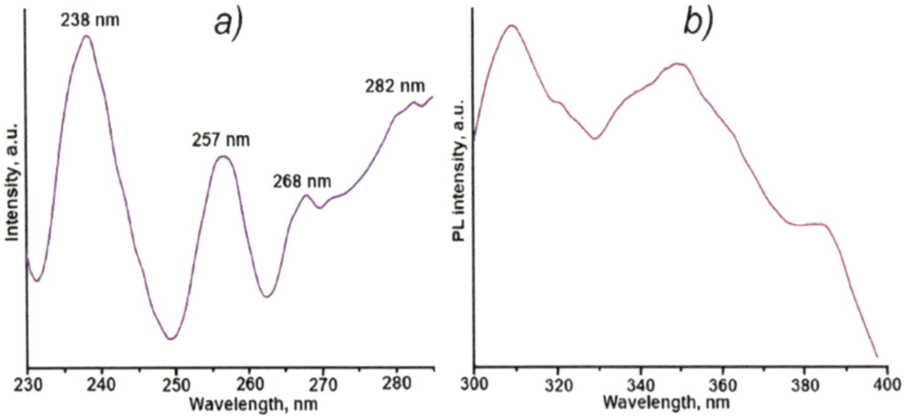


**Figure 5.**
*Luminescence excitation spectrum (wavelength of the luminescence 309 nm) of $Gd_2O_3$:$Nd^{3+}$ powder (a); photoluminescence spectrum (wavelength of the luminescence excitation 238 nm) of $Gd_2O_3$:$Nd^{3+}$ powder (b).*

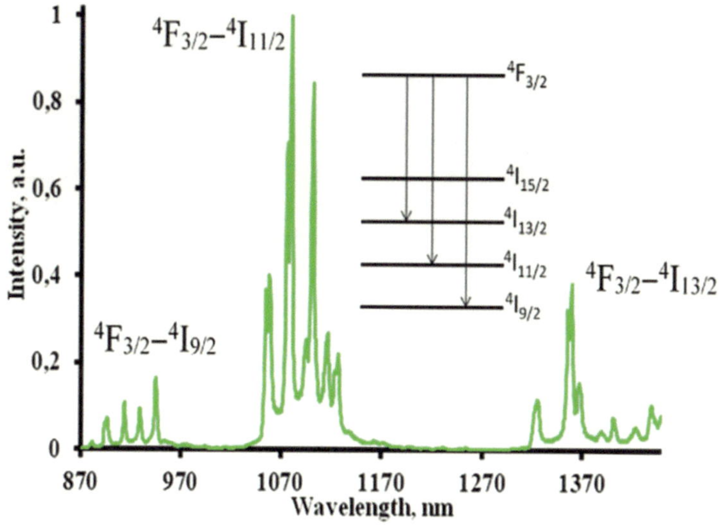


**Figure 6.**
*Photoluminescence spectrum (λex = 980 nm) of $Gd_2O_3:Nd^{3+}$ powder in the NIR region.*

### 3.6 Luminescence in IR spectral range

The intense luminescence observed in the synthesized powder in the near infrared spectral region is determined by electronic transitions of the $Nd^{3+}$ ion-activator. The photoluminescence spectrum of the synthesized $Gd_2O_3:Nd^{3+}$ powder is shown in **Figure 6**. The three groups of luminescence bands observed in the spectrum, located at wavelengths of 940, 1064, and 1360 nm, are associated with the electronic transitions of $Nd^{3+}$ions $^4F_{3/2} \rightarrow {}^4I_{9/2}$, $^4F_{3/2} \rightarrow {}^4I_{11/2}$ and $^4F_{3/2} \rightarrow {}^4I_{13/2}$, respectively. The most intense luminescence peak of neodymium ions with a maximum at $\lambda_{max}$ = 1064 nm corresponds to the electronic transition $^4F_{3/2} \rightarrow {}^4I_{11/2}$.

Analysis of the results from experiments is shown in **Figure 6**. The results presented in Ref. [28] indicate that $Gd_2O_3:Nd^{3+}$ materials synthesized by the polymer-salt process have photoluminescence spectra that are comparable to powders produced by other liquid methods. The reason for this phenomenon may be attributed to the fact that the luminescent properties, which are determined by the f-f electronic transitions of $Nd^{3+}$, are not significantly impacted by changes in the structure of their immediate surroundings.

## 4. Conclusion

Neodymium-doped gadolinium oxide nanopowders, with an average crystal size of around 30 nm, were synthesized using the citrate sol-gel method, using two organic stabilizers simultaneously. The evolution of nanoparticle structure during their formation has been studied, and the luminescent properties of the obtained nanopowders have been investigated. Citric acid and polyvinylpyrrolidone were used as modifying organic stabilizers, which acted as stabilizers of the formed $Gd_2O_3:Nd^{3+}$ nanoparticles in colloidal solutions, and also played the role of "combustible" additives in the process of heat treatment of materials. The data of infrared spectroscopy, differential-thermal, and thermogravimetric analyses showed that the formation of $Gd_2O_3$ nanoparticles starts at the raw gel stage, and the evolution process develops during drying and heat treatment of materials. The XRD data analysis reveals that the powder only consists of peaks that are typical of the cubic $Gd_2O_3$ structure. The SEM investigation demonstrates that the inclusion of polyvinylpyrrolidone has a substantial impact on the structure of the synthesized powder. It effectively separates the individual nanoparticles within the produced powders, resulting in a distinct morphology. The obtained $Gd_2O_3$:$Nd^{3+}$ nanopowders showed intense photoluminescence in the ultraviolet and near-infrared spectral regions. The study shows that PVP addition has no influence on luminescence spectra of powders synthesized. The results obtained can be used in the development of technology for the production of bulk ceramics for disc laser elements, in the creation of luminescent nanopowders for nanothermometry in medicine, as well as in the development of technology for the synthesis of various composite luminophores.

## Additional information

"This chapter is an English translation of a previously published article by the same authors: Moussaoui A, Bulyga DV, Kuzmenko NK, Ignat'ev AI, Evstropiev SK, Nikonorov NV. Sol-gel synthesis of $Gd_2O_3$:$Nd^{3+}$ nanopowders and the study of their luminescent properties. Scientific and Technical Journal of Information Technologies, Mechanics and Optics. 2021;**21**(2):198–205 (in Russian). doi: 10.17586/2226-1494-2021-21-2-198-205."

## Author details

Amir Moussaoui[1*], Dmitry V. Bulyga[1], Natalia K. Kuzmenko[1], Alexander I. Ignatev[1], Sergey K. Evstropiev[1,2] and Nikolay V. Nikonorov[1]

1 ITMO University, Saint Petersburg, Russian Federation

2 St. Petersburg State Institute of Technology (Technical University), Saint Petersburg, Russian Federation

*Address all correspondence to: am.moussaoui92@gmail.com

## References

[1] Priya R, Pandey OP, Dhoble SJ. Review on the synthesis, structural and photophysical properties of $Gd_2O_3$ phosphors for various luminescent applications. Optics and Laser Technology. 2021;**135**:106663

[2] Zhang L, Chen H, Yaohua X, Tang P, Tong Y, Ding Y. Preparation of $GdCoO_3$ by sol-gel method and its photocatalytic activity. Integrated Ferroelectrics. 2021;**219**(1):204-210

[3] Matrosova AS, Kuzmenko NK, Nikonorov NV, Aseev VA, Ananyev VA, Demidov VV, et al. Formation of $Gd_2O_3$:$Nd^{3+}$ nanocrystals in silica microcapillary preforms and hollow-core anti-resonant optical fibers. Optical Fiber Technology. 2021;**65**:102547

[4] Sharma S, Shanmugam R, Sahoo MK, Rao GR. Promoting effect of Gd2O3 in Pt-Gd2O3/C electrocatalyst for methanol oxidation reaction. Journal of the Electrochemical Society. 2022;**169**(3):034511-034511

[5] Khudiar SS, Mutlak FA H, Nayef UM. Synthesis of ZnO nanostructures by hydrothermal method deposited on porous silicon for photo-conversion application. Optik. 2021;**247**:167903

[6] Lei X, Wang H, Liwen F, Wen X, Wang J. A novel method of preparing gadolinium oxide by citric acid-assisted ultrasonic spray pyrolysis. Transactions of the Indian Institute of Metals. 2023;**76**:1615-1623

[7] Kamble C, Mane RS. Chapter 1—Introduction to wet chemical methods and metal oxide nanostructures. Metal Oxides. 2023;**1**:3-16

[8] Moussaoui A, Bulyga DV, Evstropiev SK, Ignatiev AI, Nikonorov NV, Podruhin YF, et al. Modified Pechini method by PVP addition for Nd:$Gd_2O_3$ nanophosphors fabrication. Ceramics International. 2021;**47**(24):34307-34313

[9] Dell'Era A, Catauro M. Sol–gel method applied to crystalline materials. Crystals. 2021;**11**(8):903

[10] Myasoedova TN, Kalusulingam R, Mikhailova TS. Sol-gel materials for electrochemical applications: Recent advances. Coatings. 2022;**12**(11):1625

[11] Gore SK, Jadhav SS, Tumberphale UB, Raut SD. Chapter 2—Sol-gel technology for the synthesis of metal oxide nanostructures. Metal Oxides. 2023;**1**:19-38

[12] Borlaf M, Moreno R. Colloidal sol-gel: A powerful low-temperature aqueous synthesis route of nanosized powders and suspensions. Open Ceramics. 2021;**8**:100200

[13] Michel CR, López-Contreras NL, Martínez-Preciado AH. Gas sensing properties of $Gd_2O_3$ microspheres prepared in aqueous media containing pectin. Sensors and Actuators B: Chemical. 2013;**177**:390-396

[14] Sivasamy R, Venugopal P, Mosquera E. Synthesis of $Gd_2O_3$/CdO composite by sol-gel method: Structural, morphological, optical, electrochemical and magnetic studies. Vacuum. 2020;**175**:109255

[15] Kumar RGA, Hata S, Gopchandran KG. Diethylene glycol mediated synthesis of $Gd_2O_3$:$Eu^{3+}$ nanophosphor and its Judd-Offelt analysis. Ceramics International. 2013;**39**(8):9125-9136

[16] Evstropiev SK, Vasilyev VN, Nikonorov NV, Kolobkova EV, Volkova NA, Boltenkov IA. Photoactive ZnO nanosuspension for intensification of organics contaminations decomposition. Chemical Engineering and Processing: Process Intensification. 2018;**134**:45-50

[17] Araucz K, Aurich A, Kołodyńska D. Novel multifunctional ion exchangers for metal ions removal in the presence of citric acid. Chemosphere. 2020;**251**:126331

[18] Ciriminna R, Meneguzzo F, Delisi R, Pagliaro M. Citric acid: Emerging applications of key biotechnology industrial product. Chemistry Central Journal. 2017;**11**(1):22

[19] Evstropiev SK, Soshnikov IP, Kolobkova EV, Evstropyev KS, Nikonorov NV, Khrebtov AI, et al. Polymer-salt synthesis and characterization of MgO–ZnO ceramic coatings with the high transparency in UV spectral range. Optical Materials. 2018;**82**:81-87

[20] Junaidi J, Sulistiani W, Efridahniar Y, Pratiwi I, Firdaus I, Manurung P, et al. Synthesis and characterization of Ag/$SiO_2$ nanocomposite based on rice husk silica using sol-gel method. Journal of Nano Research. 2023;**78**:31-42

[21] Buenviaje SC, Edañol YDG, Legaspi EDR, Payawan LM Jr, Usman KAS. One-pot synthesis of redispersible polymer-stabilized ZnO nanocomposites. Philippine Journal of Science. 2021;**150**(6A):1425-1435

[22] Shelemanov AA, Evstropiev SK, Karavaeva AV, Nikonorov NV, Vasilyev VN, Podruhin YF, et al. Enhanced singlet oxygen photogeneration by bactericidal ZnO–MgO–Ag nanocomposites. Materials Chemistry and Physics. 2022;**276**:125204

[23] Hao BV, Huy PT, Khiem TN, Ngueyn T, Thanh Ngan NT, Duong PH. Synthesis of $Y_2O_3$:$Eu^{3+}$ micro- and nanophosphors by sol-gel process. Journal of Physics: Conference Series. 2009;**187**:012074

[24] Moussaoui A, Bulyga D, Evstropiev S, Pchelkin G, Shurupov D, Demidov V, et al. YAG:$Ln^{3+}$ (Ln = Nd, Yb, Ce) nanocrystals for application in luminescent fiber-based temperature sensors. Optical Engineering. 2023;**62**(3):037105

[25] Tamrakar RK, Upadhyay K. Gd2O3: A Luminescent Material. In: Aide M, NakajimaT, editors. Rare Earth Elements and their Minerals. London, UK: IntechOpen; 2020. p. 186

[26] Dhananjaya N, Nagabhushana H, Nagabhushana BM, Rudraswamy B, Shivakumara C, Chakradhar RPS. Spherical and rod-like $Gd_2O_3$:$Eu^{3+}$ nanophosphors—Structural and luminescent properties. Bulletin of Materials Science. 2012;**35**(4):519-527

[27] Li QB, Lin JM, Wu JH, Lan Z, Wang JL, Wang Y, et al. Preparation of $Gd_2O_3$:$Eu^{3+}$ downconversion luminescent material and its application in dye-sensitized solar cells. Chinese Science Bulletin. 2011;**56**(28-29):3114-3118

[28] Kuzmenko NK, Evstropiev SK, Aseev VA, Danilovich DP, Nikonorov NV, Ignatiev AI, et al. Polymer-salt synthesis of $Gd_2O_3$:$Nd^{3+}$ nanophosphors. Journal of Physics: Conference Series. 2020;**1695**:012184

Chapter 5

# Upconversion Luminescent Nanoparticles and Their Biomedical Applications in Imaging

*Dalia Chávez-García and Mario Guzman*

## Abstract

Nanomaterials offer promising solutions for chemotherapy challenges, addressing issues like cytotoxicity and biocompatibility. In cancer clinical protocols, biomedical imaging is vital, providing insights into tumor morphology. Luminescent nanomaterials or nanoparticles (LNPs), particularly effective for diseases like cancer, possess controllable properties like size (usually <100 nm), surface charge, and external functionalization. LNPs interact with biological systems at systemic and cellular levels. Cellular uptake is crucial, allowing selective targeting of cancer cells through overexpressed surface receptors such as transferrin receptor (TfR), G-protein coupled receptor (GPCR), folate receptor (FR), epidermal growth factor receptor (EGFR), lectins, and low-density lipoprotein receptor (LDLR). LNPs can accumulate in subcellular compartments, playing a pivotal role in drug delivery. Studies explore LNPs' internalization into cells, investigating their potential to deliver cargoes like DNA, siRNA, miRNA, and small-molecule drugs. This review highlights the latest advancements in LNPs and their biomedical applications. Despite these promising developments, comprehensive nanotoxicological assessments are crucial for a better understanding of LNPs' behavior in biological systems, paving the way for future clinical applications.

**Keywords:** rare earth, upconversion luminescent, near-infrared, biological imaging, nanotoxicological assessment

## 1. Introduction

Nanomaterials possess unique physical properties, including magnetic, optical, mechanical, and electrical properties, which are influenced by quantum mechanics, such as electron configuration and confinement. These properties are closely linked to the size of materials produced at the nanoscale [1]. Traditionally, one-photon imaging technique that detects light emission in the visible spectra (400–700 nm) has been commonly used for in vivo optical imaging. However, in the past decade, the near-infrared (NIR-I; 700–900 nm) spectra has gained prominence due to the high light scattering coefficient of live tissues and their autofluorescence. Moreover, there is a growing interest in optical imaging in the second near-infrared (NIR-II; 1000–1700 nm) spectra, utilizing fluorescent or luminescent probes based on various materials, including inorganic quantum dots (QDs), polymer-encapsulated organic dyes, and nanoparticles [1, 2]. Researchers have developed luminescent upconversion nanoparticles (UCNPs) for NIR-I bioimaging of disease models in live mice, enhancing visualization of molecular and biological processes, as well as aiding molecular targeting-based therapies and diagnosis [3–5].

LNPs also known as phosphors or nanophosphors are solid materials capable of converting various forms of energy into electromagnetic radiation, typically in the visible range. However, they can also emit radiation in the infrared (IR), X-ray, or ultraviolet (UV) regions [6]. Particularly, UCNPs have the ability to convert NIR energy into visible light. Some types of UCNPs have a combination of Yb ions with other lanthanide ions. The $Yb^{3+}$ ion possesses an energy-level scheme characterized by a single excited 4f level of $^2f_{5/2}$ [2]. It exhibits an absorption band centered at 980 nm, attributed to the $^2F_{7/2} \rightarrow {}^2F_{5/2}$ transition. This absorption band resonates with multiple f-f transitions in upconverting lanthanide ions such as $Tm^{3+}$, $Ho^{3+}$, and $Er^{3+}$. Consequently, the $Yb^{3+}$ ion is widely employed as a sensitizer. Furthermore, the luminescence of $Yb^{3+}$ falls within the range of 960–1040 nm, making it a potential candidate for activators in NIR-II probes.

This review aims to showcase different studies involving UCNPs in bioimaging for both in vivo and in vitro applications. In the subsequent section, we will delve into the various types of luminescent nanomaterials to enhance our understanding of the underlying quantum processes involved.

## 2. Luminescent nanomaterials

Different types of luminescence phenomena exist based on the source of the excitation thus:

- Photoluminescence: This occurs when the nanomaterial is excited by electromagnetic energy, such as ultraviolet or infrared radiation.
- Cathodoluminescence: Excitation takes place when the nanomaterial is exposed to a beam of high-energy electrons.
- Triboluminescence: Mechanical energy triggers luminescence in the nanomaterial.
- Electroluminescence: Application of an electric voltage excites the nanomaterial, leading to luminescence.
- X-ray luminescence: Luminescence is induced when the nanomaterial interacts with X-rays.
- Chemiluminescence: Luminescence arises from a chemical reaction.

The luminescence of a material involves several physical processes:

1. Energy absorption: An activator ion or another ion, known as a sensitizer, absorbs the energy and transfers it to the host lattice or the activator ion.

*DOI: http://dx.doi.org/10.5772/intechopen.1005378*

2. Energy emission: The activator ion releases the absorbed energy, resulting in visible spectrum emission.

3. Relaxation to the ground state: The material returns to its ground state through non-radiative processes.

4. Interaction between luminescent centers: Energy transfer occurs between different luminescent centers.

These processes collectively contribute to the luminescent properties exhibited by nanomaterials. The host lattice plays a significant role in the luminescent process as its optical properties can vary. The ion absorbing the energy can exhibit different spectral properties depending on the host lattice, primarily influenced by factors such as covalency. Higher covalency leads to a reduction in electron interactions, causing electronic transitions between energy levels to shift to lower energy levels. For instance, the $Eu^{3+}$ ion commonly employed in LMN exhibits a higher energy absorption band when situated in the fluoride $YF_3$ compared to the covalent oxide $Y_2O_3$. Consequently, greater energy is required to excite $Eu^{3+}$ when it is in a fluoride compared to a covalent oxide [6, 7]. In this review, the focus will be on upconversion luminescent nanomaterials (UCNPs). Downconversion luminescent nanomaterials (DCNPs), on the other hand, require higher excitation energies, such as UV energy, and emit in the visible range. In contrast, UCNPs emit at higher energy wavelengths in the visible range and are excited by lower energy sources, such as near-infrared (NIR) radiation. Various types of upconversion processes exist [8]:

- Energy transfer, known as the APTE effect (Addition de Photons Par Transfers d'Energie), where ions sequentially transfer their excitation energy to another ion that can emit from a higher energy level.

- Two-step absorption, which involves a single ion with dual absorption.

- Cooperative sensitization, where two ions simultaneously transfer their excitation energy to another ion, leading to emission from the excited level.

- Cooperative luminescence, where two ions combine their excitation energy to emit a single quantum.

- Second harmonic generation (frequency doubling), which doubles the frequency of the incident light without involving any absorption transitions.

- Two-photon absorption, where two photons are simultaneously absorbed without an intermediary energy level, and emission occurs from the excited energy level.

These different processes contribute to the unique upconversion luminescence observed in nanomaterials. **Tables 1–3** showcase various types of upconversion luminescent nanomaterials (UCNPs) utilizing different host lattices, with the trivalent Yb ion serving as a sensitizer.

Research on the development of UCNPs has been ongoing for over 20 years and continues to progress. Numerous researchers have dedicated their efforts to exploring

| Host lattice | Sensitizer ion | Activator ion | Emission (s) (nm) | Reference |
|---|---|---|---|---|
| **Oxides** | | | | |
| $Y_2O_3$ | Yb | Er | 660 | [9, 10] |
| $Y_2O_3$ | Yb | Ho | 549,666 | [11, 12] |
| $Lu_2O_3$ | Yb | Er, Tm | 662,477,490 | [13, 14] |
| $Y_2O_3$ | Yb | Er, Tm | 550,660 | [15, 16] |
| $Gd_2O_3$ | Yb | Er | 600, 660 | [17, 18] |
| $La_2O_3$ | Yb | Er | 530,549,659,672 | [19] |
| **Oxysulfides** | | | | |
| $Y_2O_2S$ | Yb | Er | 520–560,650–700 | [20, 21] |
| $Y_2O_2S$ | Yb | Tm | 450–500,650,690 | [22, 23] |
| $Gd_2O_2S$ | Yb | Er | 520–580,650–700 | [23, 24] |
| **Oxyhalides** | | | | |
| GdOF | Yb | Er | 521,545,656 | [25, 26] |
| YOF | Yb | Er | 525,545,656 | [27, 28] |

**Table 1.**
*UCNP detailed with NIR excitation. Oxides, oxysulfides, and oxyhalides.*

| Host lattice | Sensitizer ion | Activator ion | Emission (s) (nm) | Reference |
|---|---|---|---|---|
| **Phosphates** | | | | |
| $LaPO_4$ | Yb | Er | 535–556 | [29, 30] |
| $LuPO_4$ | Yb | Tm | 476 | [31] |
| Molybdates | | | | |
| $La_2(MoO_4)_3$ | Yb | Er | 519,541 | [32, 33] |
| $La2(MoO_4)_3$ | Yb | Tm | 472,647 | [34] |
| **Tungstates** | | | | |
| $NaY(WO4)_2$ | Yb | Er | 526,553,660 | [35, 36] |
| $NaY(WO4)_2$ | Yb | Tm | 476,647 | [12, 35, 36] |
| **Gallates** | | | | |
| $Gd_3Ga_5O_{12}$ | Yb | Tm | 454,484,640–680 | [37, 38] |
| **Vanadate** | | | | |
| $YVO_4$ | Yb | Er | N/A | [39, 40] |

**Table 2.**
*UCNP detailed with NIR excitation. Phosphates, molybdates, tungstates, gallates, and vanadates.*

and identifying UCNPs with optimal luminescent properties in specific host lattices. **Tables 1–3** present a comprehensive overview of the various types of UCNPs created using different host lattices and dopant ions.

The intensity of luminescence in different combinations of ions is influenced by both the host lattice and ion doping. It is important to note that high doping

| Host lattice | Sensitizer ion | Activator ion | Emission (s) (nanometers) | Reference |
|---|---|---|---|---|
| Fluorides | | | | |
| $LaF_3$ | Yb | Er | 521,545,659 | [41, 42] |
| $LaF_3$ | Yb | Tm | 475,698,800 | [19, 43] |
| $LaF_3$ | Yb | Ho | 541,643 | [44, 45] |
| $YF_3$ | Yb | Er | 411,526,552,664 | [44, 46] |
| $YF_3$ | Yb | Tm | 347,636,454,477 | [47, 48] |
| $LuF_3$ | Yb | Tm | 481 | [49, 50] |
| $NaYF_4$ | Yb | Er | 525,547,660 | [27, 51, 52] |
| $NaYF_4$ | Yb | Tm | 450,476,800 | [53, 54] |
| $NaYF_4$ | Yb | Ho | 541 | [55] |
| $LiYF_4$ | Yb | Tm | 361,450,479,647 | [56] |
| $NaGdF_4$ | Yb | Ho | 541,647,751 | [57] |
| $KY_3F_{10}$ | Yb | Er | 522,545,656 | [58, 59] |
| $KGd_2F_7$ | Yb | Er | 525,552,666 | [59] |
| $BaYF_5$ | Yb | Tm | 475,650,800 | [60–62] |

**Table 3.**
*UCNP detailed with NIR excitation for fluorides.*

does not necessarily guarantee high luminescence. The energy transfer process between ions is governed by several factors, such as the crystal structure of the host lattice, which determines the distance and spatial arrangement of the doping ions. Therefore, the host lattice should have a low phonon energy and a close lattice match with the dopant ions. These characteristics help minimize non-radiative losses and maximize radiative emission [63–65]. The successful utilization of UCNPs in biomedicine relies on careful selection of nanoparticles, considering factors such as size, shape, ease of functionalization, and tendency to agglomerate, among others. These factors are crucial for ensuring optimal performance and compatibility with biomedical applications.

A wide variety of luminescent nanoparticles (NPs) can be utilized for biomedical applications, distinguished by the chemical nature of their host lattice. Different systems include sulfides (ZnS, CaS, BaS, SrS) [63], oxides (ZnO, $Fe_3O_4$) [64], and lanthanides oxides [65]. These systems exhibit strong emission spectra ranging from 450 to 700 nm, achieved by incorporating trivalent lanthanide ions such as Yb, Er, and Tm. More specific details regarding these combinations can be found in **Table 1**.

**Table 2** highlights less commonly used host lattices, including phosphates, molybdates, tungstates, gallates, and vanadates. It is worth noting that oxides, in general, tend to exhibit greater stability. However, molybdate hosts have a small optical band gap, which can result in thermal quenching, limiting their suitability in certain applications. **Table 3** focuses on fluorides, such as CaF [66] and lanthanides fluorides [67]. These types of host lattices are popular due to their advantageous properties. However, one disadvantage is that it can be more challenging to obtain nanocrystals with sizes smaller than 50 nm using these host lattices.

Several of the mentioned systems offer not only luminescence but also additional exciting features. For instance, $Fe_3O_4$ nanoparticles exhibit magnetic properties [64], making them highly valuable for more efficient targeting of cancer cells. Gold nanoparticles supported on $TiO_2$ show a photo-thermal response [68], allowing for applications in targeted photothermal therapy. $CeO_2$ nanoparticles exhibit antimicrobial behavior providing potential applications in combating microbial infections [69]. Additionally, inorganic nanoparticles with high surface area and porosity, such as those based on certain materials, can improve the ability to carry drugs [70]. The chemical nature of inorganic nanoparticles directly influences their functionalization, enabling specific modifications for desired applications. Furthermore, the crystal properties of these inorganic nanoparticles are crucial for understanding and controlling their luminescent behavior in synthesized materials. A comprehensive understanding of both the chemical and crystal properties is vital in harnessing the full potential of these materials for various applications.

## 2.1 Synthesis of the nanoparticles

Achieving high-quality UCNPs requires careful control over their composition, size, shape, and crystalline phase. The synthesis process plays a critical role in attaining these desired characteristics. The most popular synthesis methods for UCNPs are based on the "bottom-up" approach, commonly known as wet chemistry. This approach involves chemical reactions in a liquid solution to form nanocrystals. Wet chemistry methods offer advantages in producing uniform nanocrystals, allowing for precise control over particle size and minimizing agglomeration.

By employing the bottom-up approach, researchers can carefully manipulate reaction parameters such as precursor concentrations, reaction time, temperature, and choice of solvent to achieve the desired properties of UCNPs. This control over the synthesis process enables the production of UCNPs with tailored characteristics, facilitating subsequent functionalization and integration into specific applications [71]. The LNPs, in general, can be synthesized in the same way, no matter if they are DCNP or UCNP.

### *2.1.1 Sol-gel method*

The sol-gel process is a synthesis method that involves the hydrolysis and polycondensation of metal alkoxides (or halides) as precursors. This process is conducted at relatively low temperatures; however, to improve the crystallinity of the resulting nanomaterials, calcination at high temperatures is often required. The high luminescence efficiency of the nanoparticles can be attributed to this post-calcination step.

Several researchers have explored the sol-gel process for the synthesis of luminescent nanoparticles. For example, Taxak [72] utilized tartaric acid-assisted $Y_2O_3{:}Eu^{3+}$ DCNPs using this method. Talane et al. [73] prepared $TiO_2{:}Er^{3+}$ UCNPs through the sol-gel process. Chavez et al. [10] also employed the sol-gel method to produce tartaric acid-assisted UCNPs with two different host lattices: $Y_2O_3$ and $Gd_2O_3$. In their study, $Y(NO_3)_3$ and $Gd(NO_3)_3$ were used as precursors, along with doping precursors $Yb(NO_3)_3$ and $Er(NO_3)_3$. The $Y_2O_3$ nanoparticles required higher calcination temperatures of up to 1200°C.

However, one drawback of the sol-gel process is the need for higher calcination temperatures, which can be a limitation in terms of energy requirements and potential impact on the material or substrate being used.

### *2.1.2 Thermal decomposition*

Thermal decomposition is another synthesis method that can yield high-quality UCNPs with controlled size and desirable shapes [74]. This method involves two steps. In the first step, organic precursors are dissolved in high-boiling point organic solvents along with surfactants. In the second step, the precursors are decomposed at elevated temperatures. Organic acid compounds such as trifluoroacetate, acetoacetate, and oleate, among others, can serve as precursors in this process.

Boyer et al. [75] demonstrated the synthesis of $NaYF_4$ UCNPs co-doped with Yb/Er or Yb/Tm using thermal decomposition with metal trifluoroacetate precursors in the presence of oleic acid and octadecene. Oleic acid not only acted as a solvent but also as a passivating ligand to prevent agglomeration of the UCNPs.

It is important to note that there are certain disadvantages associated with this method. The use of toxic or expensive precursors can be a limitation in some cases, and it is necessary to handle them with care. Additionally, the cost of the process itself can be relatively high due to the requirements of elevated temperatures and the use of specialized equipment.

### *2.1.3 Hydrothermal method*

The hydrothermal method is a widely used technique to produce luminescent nanoparticles with high crystallinity. This method involves using solvents, either hydrophilic or lyophilic, at high temperatures inside autoclaves to create autogenic high pressure. The elevated temperature and pressure promote the solubility and reordering of precursors, resulting in the formation of high-quality nanoparticles.

The hydrothermal method typically operates at relatively low reaction temperatures, allowing for precise control over the morphology of the nanoparticles. Lin et al. developed an effective method for producing nanoparticles with different shapes based on rare earth elements. They discovered that the addition of trisodium citrate as an organic additive, the choice of fluoride source, and the pH value of the initial solution greatly influenced the shape of the nanoparticles [76].

However, it is worth noting that there are limitations to this method. Nanoparticles produced via hydrothermal synthesis generally cannot be smaller than 10 nm in size. Additionally, some hydrothermal reactions may require prolonged heating periods to complete the synthesis.

### *2.1.4 Coprecipitation method*

Coprecipitation is indeed a commonly used method for synthesizing ultrasmall NPs. It offers advantages such as simplicity, low cost, and does not require complex equipment or procedures. In coprecipitation, the simultaneous precipitation of multiple components allows for the formation of nanoparticles.

Typically, a subsequent heat treatment step is required to improve the crystallinity of the synthesized NPs. This heat treatment is commonly performed at temperatures ranging between 400 and 600°C.

In a study by Yi et al. [77], they synthesized $LaF_3{:}Yb^{3+}$, $Er^{3+}/Ho^{3+}/Tm^{3+}$ UCNPs with a very small particle size of 5.4 ± 0.9 nm. They utilized ethylenediamine tetraacetic acid as a chelating reagent in the coprecipitation process. Overall, coprecipitation is a versatile and straightforward method for producing ultrasmall nanoparticles, and subsequent heat treatment can enhance their crystallinity and properties.

### *2.1.5 Combustion synthesis*

Combustion synthesis is a highly exothermic reaction method, typically performed within a temperature range of 500 to 3000°C. Despite its tendency to cause agglomeration during the process, several authors have successfully utilized this method for the synthesis of upconversion nanoparticles (UCNPs).

In a study by Pendey et al. [78], $Y_2O_3$ co-doped with $Ho^{3+}$ and $Yb^{3+}$ was synthesized using combustion synthesis, resulting in high green emissions. Although agglomeration can occur during combustion synthesis, these methods remain widely used for the production of UCNPs in biomedical applications. Researchers often personalize and refine their synthesis methods to achieve spheroidal nanoparticles with reduced agglomeration. Surfactants are sometimes incorporated into the process to improve the quality and shape of the nanoparticles. These efforts aim to enhance the overall performance and applicability of the synthesized UCNPs.

The following **Table 4** presents a summary of the various methods explained. These are not the only methods that may exist; we are presenting the most common ones used by researchers.

## 2.2 Functionalization of nanoparticles

The functionalization of nanoparticles plays a crucial role in the development of materials for cancer treatment. This process involves modifying the surface of the nanoparticles to make them compatible with biological systems and enable specific targeting of cancer cells.

To achieve this, the nanoparticles are made biocompatible by attaching a chemical group that facilitates the binding of various ligands. These ligands are designed to specifically bind to overexpressed receptors on cancer cells, such as the folate receptor (FR), transferrin receptor (TfR), epidermal growth factor receptor (EGFR),

| Method | Advantages | Disadvantages | References |
|---|---|---|---|
| Sol–gel | It is performed at relatively low temperatures | High calcination temperatures | [72, 73] |
| Thermal decomposition | It can result in high-quality UCNPs with precise control over their size and shapes | The utilization of toxic or costly precursors entails careful handling precautions and incurs elevated expenses | [74, 75] |
| Hydrothermal | It can produce high crystallinity | LNPs cannot be smaller than 10 nm in size. Certain hydrothermal reactions may necessitate prolonged heating periods to achieve complete synthesis | [76] |
| Coprecipitation | Simple, low cost, and does not require complex equipment | Heat treatment is required to improve the crystallinity | [77] |
| Combustion synthesis | Fast and highly exothermic reaction method | It has a tendency to cause agglomeration of the LNPs during the process | [78] |

**Table 4.**
*Synthesis of LNPs.*

G-protein coupled receptor (GPCR), low-density lipoprotein receptor (LDLR), and lectins, among others. **Figure 1** illustrates the basic concept of UCNPs and the functionalization process that enables them to target specific cells. The first step involves the synthesis of UCNPs using the upconversion process. Then, the nanoparticles are functionalized to make them biocompatible and capable of binding to specific ligands. Finally, the functionalized UCNPs can bind to the desired cancer cells, allowing for targeted delivery or imaging applications. This functionalization process is essential for enhancing the selectivity and effectiveness of cancer treatment by enabling nanoparticles to specifically interact with cancer cells while minimizing interactions with healthy cells.

**Figure 1** illustrates the fundamental procedure required to functionalize UCNPs for targeting specific cells. For example, folic acid (FA) is one of the chemical groups commonly used to modify the surface of nanoparticles due to the overexpression of folate receptors (FR) in many human cancer cells [79, 80]. The presence of FR makes folic acid an attractive ligand for targeted drug delivery to cancer cells.

Polyethylene glycol (PEG) is another widely used chemical group for nanoparticle functionalization. PEG functionalization enhances the biocompatibility of nanoparticles due to the hydrophilic nature of the PEG molecule [81]. The presence of PEG on the nanoparticle surface reduces non-specific interactions and helps to increase circulation time in the body.

The size of nanoparticles also plays a significant role in their biodistribution and targeting efficiency. Cheng et al. [82] conducted research on *in vivo* targeted drug delivery using nanoparticles for prostate tumors. They found that the nanoparticle size had a significant impact on the biodistribution of targeted and non-targeted nanoparticles in an organ-specific manner. This study highlights the importance of nanoparticle size in achieving effective targeted drug delivery.

Overall, the functionalization of nanoparticles with chemical groups like FA and PEG enables targeted drug delivery and enhances their biocompatibility, making them suitable for biomedical applications. Manson et al. [83] conducted a study on

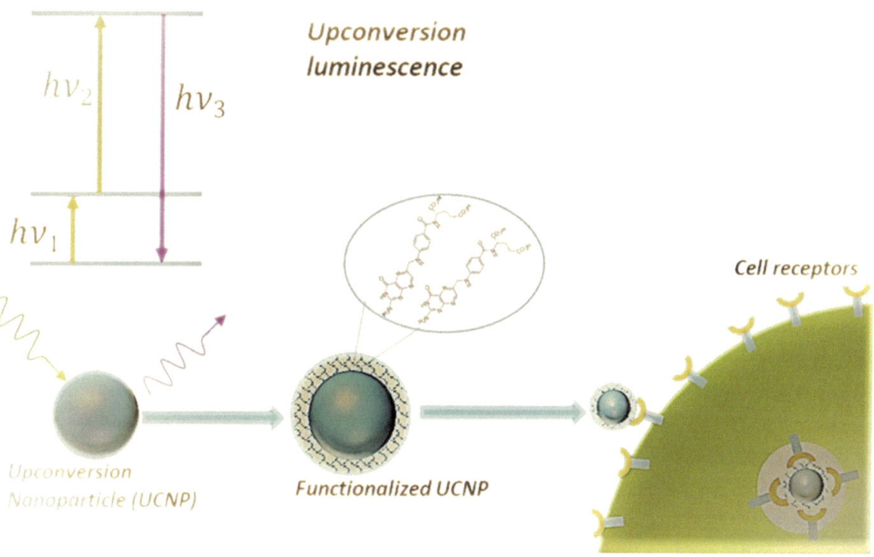


**Figure 1.**
*Image shows the basic procedure needed to functionalize UCNPs to target specific cells.*

the capping density of PEG on gold nanoparticles (AuNPs) and its impact on the surface properties of the nanoparticles in various solutions. The study revealed that the PEG-coated nanoparticles exhibited significantly increased stability, facilitating their production in different media. This enhanced stability is essential for ensuring the nanoparticles' functionality and performance in various applications.

In another study by Maldiney et al. [84], the effect of optical detection using PEGylated lanthanide-doped luminescent nanoparticles (NPs) emitting in the near-infrared (NIR) window was investigated in healthy and tumor-bearing mice. The researchers found that the tissue distribution of the nanoparticles was highly dependent on the surface coverage of PEG as well as the core diameter of the nanoparticles. They concluded that PEG coverage led to a homogeneous and well-distributed presence of highly stable particles throughout the animal, highlighting the importance of PEGylation for achieving effective and stable distribution of nanoparticles in biological systems.

These studies emphasize the significance of PEG functionalization in enhancing the stability, production, and distribution of nanoparticles in various media and biological systems. PEGylation plays a crucial role in ensuring the biocompatibility, stability, and targeted delivery capabilities of nanoparticles, making them valuable tools in biomedical applications.

Silica ($SiO_2$) coating is another effective method for functionalizing nanoparticles. Wang et al. [41] conducted a study on UCNPs coated with silica (LaF3: Yb, Er@$SiO_2$) in zebrafish embryos. They found that a high dose of fluoride UCNPs (> 200 μg/mL) could lead to malformations, delayed hatching rate, and impaired embryonic and larval development. The study highlights the importance of carefully considering the dosage and potential toxicity of nanoparticles, even when they are functionalized with silica coatings. While functionalization can improve the biocompatibility and stability of nanoparticles, it is crucial to ensure that the dosage used does not have adverse effects on biological systems. This underscores the importance of conducting thorough toxicity assessments and dose optimization studies when utilizing functionalized nanoparticles in biological applications.

Polyethyleneimine (PEI) is indeed a commonly used method for functionalizing nanoparticles. When combined with small-interfering RNA (siRNA), it offers a powerful strategy for targeted therapy in various diseases. siRNA is a synthetic double-stranded RNA molecule consisting of approximately 21 nucleotides. It is designed to trigger a specific and sequential degradation of target mRNA through the RNA interference (RNAi) mechanism. This functionalization approach provides a potential strategy for silencing mutated or defective genes that contribute to various human diseases. Buchman et al. [85] developed a new method for the covalent attachment of PEI for siRNA delivery. They demonstrated that nanoparticles functionalized with this approach exhibited thermodynamic stability, efficient binding to siRNA molecules, and sustained silencing effects even after storage for at least 1 year. This research highlights the potential of PEI-based functionalization for effective and long-term delivery of siRNA, enabling targeted gene silencing for therapeutic purposes.

Polyvinylpyrrolidone (PVP) and polyacrylic acid (PAA) are indeed commonly used for the functionalization of nanoparticles. Wang et al. conducted research on the functionalization of upconversion nanoparticles (UCNPs) $NaYF_4$:Yb, Er with various coatings including PEG, PVP, PAA, and PEI with -COOH and -$NH_2$ groups [27, 53]. These functionalized nanoparticles have shown great potential for applications in biomedicine. In another study by Liu et al. [86], PAA functionalization of UCNPs with -COOH and -$NH_2$ groups was employed for immunoassays. The functionalized UCNPs demonstrated high sensitivity for biosensing and exhibited high contrast for bioimaging applications.

The use of PVP, PAA, and other functionalization agents allows for the modification of nanoparticles' surface properties and facilitates their interactions with biological systems. These functionalized nanoparticles have shown promise in various biomedical applications, including drug delivery, biosensing, and bioimaging, due to their enhanced biocompatibility and tailored surface properties.

In the study by Yang [87], NaREF4 nanoparticles (NPs) co-doped with Nd–Lu and Y were obtained using a phase-transition process. This process involved the use of sodium dodecylsulfate (SDS) as an amphiphilic surfactant within the same reaction system. As a result, hollow-structured NPs were generated in situ. These hollow-structured NPs have unique properties and can be easily synthesized using electron-beam lithography on solid NPs. They exhibited good biocompatibility, making them suitable for biomedical applications. In the study, the NPs were tested for drug storage and delivery, as well as for cancer therapy drug carrier studies. The hollow structure of these NPs provides opportunities for drug encapsulation, controlled release, and targeted delivery. This research contributes to the exploration of novel nanomaterials for drug delivery systems and holds potential for advancements in cancer therapy.

Folic acid is an essential vitamin involved in various metabolic reactions, including the synthesis of purine and pyrimidine bases. Due to the high demand for these bases in rapidly proliferating tumor cells, FA is required in larger quantities by cancer cells. Moreover, several types of cancer cells, such as ovarian and uterine cancer cells, exhibit overexpression of folate receptors at higher rates compared to normal cells. This makes FA an excellent targeting molecule for cancer cells in both in vitro and in vivo applications.

Suvorov et al. reported a non-invasive method using FA-functionalized nanoparticles (NPs) conjugated with photosensitizers (PS) for photodynamic therapy (PDT) of cancer. In this approach, the PS is taken up by the tumor, and upon local irradiation with light of a specific wavelength, active oxygen forms, including singlet oxygen, are generated. These active oxygen species cause cell death in the tumor cells and surrounding tissue, leading to therapeutic effects [88, 89]. Overall, the use of FA-functionalized NPs holds great promise for targeted drug delivery and therapeutic applications, offering the potential for enhanced cancer treatment outcomes while minimizing off-target effects on healthy cells.

In a study conducted by Xie et al. [90], the self-assembly abilities of Brome mosaic virus (BMV) capsid proteins were investigated in the aprotic polar solvent dimethyl sulfoxide (DMSO), which differs from the typical aqueous buffers used in previous studies. The researchers observed that the BMV capsid proteins retained their ability to self-assemble in DMSO, enabling the encapsulation of nanoparticles and dye molecules that are more soluble in organic solvents, such as β-NaYF4-based UCNPs and BODIPY dye. Surprisingly, the assembly process demonstrated robustness and adaptability across a wide range of DMSO concentrations. Particularly interesting was the finding that cargos with poor stability in aqueous solutions could be easily encapsulated at high DMSO concentrations and subsequently transferred to aqueous solvents, where they maintained stability and functionality for an extended period of several months. This study underscores the potential of utilizing alternative solvents, such as DMSO, for virus assembly and encapsulation, opening up new avenues for the development of functional nanomaterials.

Ensuring the biocompatibility of functionalized NPs is a critical consideration for their use in biomedical applications. Additionally, minimizing particle aggregation is important to maintain the circulation of NPs within the human body and their efficient delivery to targeted cancer cells. Numerous studies have focused on achieving efficient functionalization of NPs, resulting in successful uptake of NPs by cells and improved biocompatibility.

Indeed, the coating and functionalization of nanoparticles can pose challenges such as agglomeration. To overcome this issue, surfactants are commonly employed to reduce agglomeration and achieve adequate coating, ensuring the biocompatibility of the nanoparticles and facilitating their successful penetration into cells. Surfactants act by stabilizing the nanoparticles, preventing them from clumping together, and promoting their dispersion.

In addition to surfactants, mechanical methods such as sonication are often utilized to enhance the dispersion of nanoparticles before functionalization. Sonication involves the application of high-frequency sound waves, which create cavitation bubbles that implode, exerting shear forces and promoting the breakup of particle aggregates. This mechanical treatment helps to disperse the nanoparticles more effectively, enhancing their stability and facilitating subsequent functionalization processes.

By utilizing surfactants and employing mechanical methods like sonication, researchers aim to minimize agglomeration and achieve better dispersion and coating of nanoparticles. These strategies contribute to improving the biocompatibility of nanoparticles and their ability to interact with target cells, ultimately enhancing their potential for biomedical applications.

## 3. Bioimaging with nanomaterials

Bioimaging is a powerful technique that enables the visualization of various biological structures and their functions with high resolution. It offers a level of detail that surpasses traditional imaging methods such as X-ray or magnetic resonance imaging (MRI). Bioimaging can be conducted on living organisms, providing insights into dynamic processes and interactions within the body, or on *in vitro* cell cultures to study cellular behaviors and molecular processes.

In the context of *in vivo* bioimaging, it allows researchers to non-invasively observe and track biological phenomena in real time within living animals. This includes imaging specific tissues or organs, studying the progression of diseases, monitoring therapeutic responses, and exploring the biodistribution and pharmacokinetics of drugs or nanoparticles.

For *in vitro* bioimaging, cell cultures can be examined under various microscopy techniques to investigate cellular structures, protein localization, gene expression, and cellular responses to external stimuli. This enables researchers to gain a deeper understanding of cellular processes, cell-cell interactions, and the effects of various interventions.

Bioimaging techniques encompass a wide range of methods, including fluorescence microscopy, confocal microscopy, multiphoton microscopy, positron emission tomography (PET), single-photon emission computed tomography (SPECT), and more. Each technique offers specific advantages and capabilities, allowing researchers to tailor their imaging approach to the specific biological question or experimental setup [91].

In summary, bioimaging plays a crucial role in advancing our understanding of biological systems, both in vivo and *in vitro*. It provides detailed visualizations of structures and processes, enabling researchers to unravel the complexities of biology and contribute to advancements in fields such as medicine, drug development, and basic biological research [87, 92, 93].

### 3.1 *In vitro* cellular imaging

*In vitro* cellular imaging provides an intuitive visualization of physiological processes and interactions among cells, proteins, and molecules [94]. Researchers have utilized laser scanning upconversion luminescence (UCL) microscopy systems, such as the one developed by Yu and colleagues [95], for cellular imaging with UCNPs. The unique anti-Stokes emission modality of UCNPs, which differs from single-photon and two-photon emissions, presents a challenge for cellular imaging due to out-of-focus upconversion emissions. To overcome this, a reverse excitation dichroic mirror and confocal pinhole technique were employed in the microscopy system.

The common incubation concentration of UCNPs with cells for cellular imaging typically ranges between 10 and 200 μg/mL. Hu et al. developed cellular imaging using PEG-OH $LaF_3$: Yb, Ho UCNPs [45]. They achieved good cell membrane permeability with KB cells, enabling the UCNPs to serve as bioimaging probes. The UCNPs exhibited strong intracellular upconversion emissions within the wavelength range of 545 to 645 nm.

The effect of coated UCNPs on cellular uptake and imaging has been studied by various researchers. Jin et al. [96] investigated cubic $NaYF_4$: Yb, Er/Tm UCNPs coated with different polymers for cellular imaging They synthesized PVP-UCNPs, PAA-UCNPs, and PEI-UCNPs using polyvinylpyrrolidone (PVP), polyacrylic acid (PAA), and polyethyleneimine (PEI), respectively. Based on imaging results, they found that PEI-UCNPs exhibited more effective intracellular uptake. This finding aligns with a report by Tsang et al. who obtained similar results in their study with HeLa cells. The significant cell uptake of PEI-UCNPs is attributed to the electrostatic attraction between the amino groups ($-NH_2$) on PEI and the cell membrane [97].

These studies highlight the importance of surface coatings and their impact on cellular uptake and imaging performance of UCNPs. The choice of coating material can significantly influence the cellular uptake efficiency, allowing for enhanced imaging capabilities and better understanding of cellular processes.

**Figure 2** depicts an example of how the in vitro cellular imaging appears with confocal microscopy of cervical (HeLa) and breast (MCF-7) adenocarcinoma cell lines [98].

### 3.2 *In vivo* tumor imaging

*In vivo* tumor imaging is crucial for understanding the mechanism of drug function. To address the limitations of UCNP applicability in in vivo imaging, researchers have explored the use of UCNPs doped with $Tm^{3+}$ and $Er^{3+}$ ions. These UCNPs offer better performance for deep tissue imaging. De la Rosa et al. highlighted that the optimal excitation wavelength for UCNP, around 980 nm, coincides with the absorption band of water, leading to potential tissue heating. However, when UCNP is e Raman micro-spectroscopy excited at 915 nm wavelength, optical excitation of $Yb^{3+}$ can still occur without interference from water [99–101].

UCNPs have several advantages for *in vivo* imaging. They can be designed to emit fluorescence under specific excitation wavelengths, allowing for targeted activation and detection of disease-specific tissues. Xiong et al. [102] functionalized UCNPs with FA and demonstrated significant upconversion luminescence signals in mice tumors after intravenous injection.

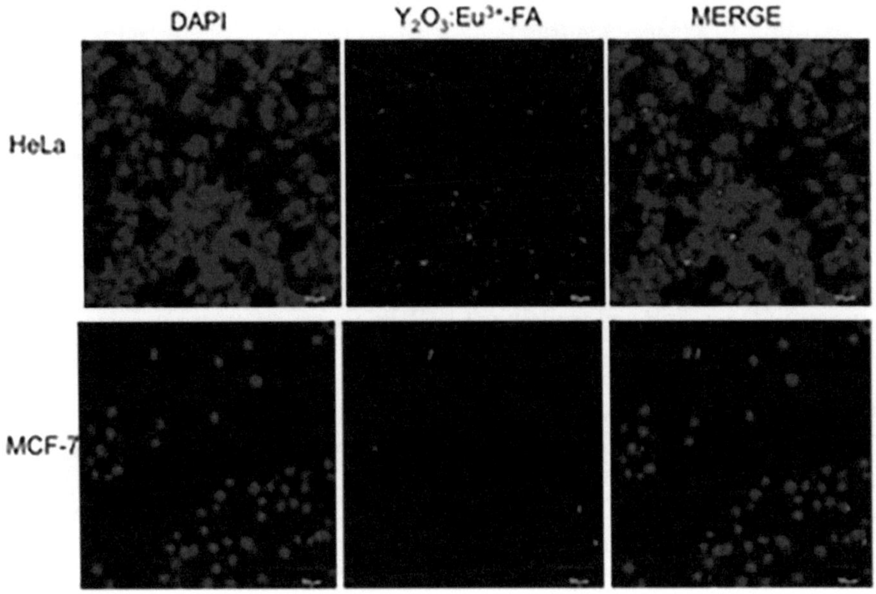


**Figure 2.**
*Cellular localization of fluorescence Y2O3:$Eu^{3+}$-FA nanoparticles in cancer cells. Cervical (HeLa) and breast (MCF-7) adenocarcinoma cell lines incubated with LNPs-FA (red). DAPI depicts nuclear staining in blue [98].*

Gadolinium-based UCNPs, where $Gd^{3+}$ ions are incorporated into the host lattice, offer both magnetic resonance imaging (MRI) and luminescence contrast, enabling multimodal imaging. Similarly, UCNP $NaLuF_4$: Yb, Tm can act as dual contrast agents for computerized tomography (CT) and upconversion luminescence in lymphatic imaging. Zeng et al. [103] employed tri-modal imaging combining radionuclide labeling with MRI- or CT-capable UCNP PEG-$BaGdF_5$: Yb/Er for in vivo tumor diagnosis.

One challenge of UCNP is their relatively low brightness, which can be attributed to narrow excitation bands, low absorption cross-sections, and surface-induced quenching effects. Researchers have addressed these issues by manipulating the host structure through the addition of dopant ions or by modifying the NP surface with molecules or materials that enhance light absorption, such as dyes and plasmonic materials [4]. High-resolution in vivo imaging is desirable for achieving detailed spatial resolution. Hilderbrand et al. [104] studied the use of $Y_2O_3$: Yb, Er UCNPs for in vivo vascular imaging in mice, demonstrating clear images of vessels and long circulation of UCNPs for up to 2 hours post-injection.

UCNPs offer significant advantages in bioimaging for both *in vivo* and *in vitro* applications. They provide tissue autofluorescence-free imaging and improved spatial resolution due to reduced scattering at longer wavelengths. Ongoing research efforts aim to address the challenges and improve the utilization of UCNPs in bioimaging [105].

## 4. Cellular uptake and toxicology of nanoparticles

### 4.1 Toxicology of nanoparticles

With the increased use of UCNPs, concerns have been raised regarding their potential impact on human health and biological systems. Consequently, numerous studies have been conducted to assess the uptake of UCNPs into cells or tissues and evaluate their potential toxicity.

Several assays are commonly employed to assess the cytotoxicity of NPs in cells, such as the MTT assay. This assay measures cell viability by assessing the ability of cells to convert the yellow tetrazolium salt (MTT) into a purple formazan product through mitochondrial activity. Positive and negative controls are used to determine total cell death and cell survival, respectively. Cells are incubated with UCNPs at various concentrations for specific incubation periods, and the cell viability is then measured using an ELISA reader or a similar method that quantifies the color change [106]. For example, Shan et al. [107] performed a biofunctionalization study on $NaYF_4$: Yb, Er UCNPs. They coated the UCNPs with a layer of $SiO_2$ and introduced amino and carboxyl groups. The UCNPs were incubated with cells at a concentration of 1 mg/mL for 9 days. Their results showed that UCNPs functionalized with amino groups exhibited 96.2% cell viability, while those functionalized with carboxyl groups exhibited 92.8% cell viability, indicating msinimal toxicity.

In addition to the MTT assay, other assays are employed to evaluate potential toxic effects of UCNPs, such as assessing reactive oxygen species (ROS) generation, apoptosis, necrosis cell death assays, and genotoxicity using the comet assay. These assays provide valuable insights into the potential biological impact of UCNPs and help assess their safety profile.

Reactive oxygen species (ROS) are highly reactive molecules that can cause oxidative damage to biological macromolecules, leading to functional impairment. ROS are generated during cellular respiration and play important roles in various physiological processes, including cell development, differentiation, oxidative stress, and cell death. The formation of ROS by nanoparticles is one of the mechanisms through which nanoparticles can exert toxicity [108].

Many studies have investigated the generation of ROS by nanoparticles as a means to evaluate their biocompatibility. One commonly used method for detecting ROS and reactive nitrogen species (RNS) is the reaction of 2,2,6,6-tetramethylpiperidine (TEMP) with superoxide radicals ($O^{2-}$), which can be detected using X-band electron paramagnetic resonance (EPR). However, this method can be costly compared to using fluorescent probe molecules. Another limitation of the EPR method is its inability to react with a wide range of reactive species. In addition to the EPR method, there are other techniques available for detecting and quantifying ROS, such as fluorescence-based probes. These probes can selectively react with specific ROS species, providing a more comprehensive assessment of ROS generation and their potential effects [109].

Overall, investigating the generation of ROS by nanoparticles is an important aspect of assessing their toxicity and biocompatibility. The choice of detection method depends on various factors, including the specific ROS species of interest, the sensitivity required, and the cost-effectiveness of the method.

Flow cytometry is a powerful technique for analyzing cell health and assessing compound cytotoxicity. It provides a detailed and comprehensive analysis of various cellular parameters, including cell viability, apoptosis, and necrosis. Flow cytometry utilizes fluorescently labeled probes and antibodies to measure specific cellular markers and characteristics [110].

Cell death can be assessed by flow cytometry through the detection of apoptosis and necrosis markers. Apoptosis is a programmed cell death process characterized by specific morphological and biochemical changes, such as membrane blebbing, DNA fragmentation, and caspase activation. Necrosis, on the other hand, is a form of cell death typically associated with cellular damage and disruption, leading to the release of cellular contents. To discriminate between apoptotic and necrotic cell death, specific fluorescent markers can be used. For example, Annexin V is commonly used to detect early apoptotic cells by binding to phosphatidylserine exposed on the outer leaflet of the plasma membrane. Propidium iodide (PI) or 7-aminoactinomycin D (7-AAD) can be used to assess necrotic cells by detecting the loss of plasma membrane integrity [111–113].

In addition to flow cytometry, Raman micro-spectroscopy has emerged as a non-invasive and label-free technique for studying cell death processes. Raman spectroscopy measures the vibrational frequencies of molecular bonds, providing information about the biochemical composition and structural changes within cells. By analyzing Raman spectral shifts associated with specific cellular components, such as proteins, lipids, and nucleic acids, researchers can differentiate between different modes of cell death, including apoptosis and necrosis [114, 115].

The study conducted by Brauchle and et al. [116] aimed to investigate induced apoptosis and heat-triggered necrosis in Saos-2 and SW-1353 cells using Raman micro-spectroscopy at room temperature. Raman micro-spectroscopy is a label-free technique that allows for the analysis of biochemical composition and structural changes within cells. In the study, the researchers exposed the cells to specific stimuli to induce apoptosis and heat-triggered necrosis. They then performed Raman micro-spectroscopy measurements on the cells to analyze the resulting spectral shifts associated with these different cell death processes. By examining the Raman spectra of the cells, the researchers were able to identify distinct biochemical signatures associated with apoptosis and necrosis. Apoptotic cells exhibited characteristic spectral changes, such as alterations in protein conformation and nucleic acid content. On the other hand, heat-triggered necrotic cells displayed different spectral features, including changes in lipid composition and membrane integrity. The use of Raman micro-spectroscopy in this study allowed for the discrimination and identification of different modes of cell death, providing valuable insights into the biochemical changes associated with apoptosis and necrosis. This non-invasive and label-free technique offers the advantage of studying complex cell-death-related processes in a continuous and real-time manner, without the need for exogenous labels or probes. Overall, the study by Brauchle and colleagues demonstrated the utility of Raman micro-spectroscopy in elucidating the biochemical characteristics and dynamics of cell death processes, contributing to a deeper understanding of these important biological events.

The combination of Raman micro-spectroscopy with other techniques, such as fluorescent microscopy, allows for a more comprehensive understanding of complex cell-death-related processes. This approach enables continuous and non-invasive monitoring of cell death events, providing valuable insights into the underlying mechanisms and dynamics of cell response to cytotoxic stimuli.

The genotoxicity testing is crucial for evaluating the potential harmful effects of chemicals on the genetic material of cells. The comet assay, also known as the single-cell gel electrophoresis assay, is a widely used technique to assess DNA damage and genotoxicity. It involves exposing cells to a test substance and then subjecting them to electrophoresis, which causes fragmented DNA to migrate away from the nucleus, creating a "comet" shape. The comet assay allows researchers to measure various types of DNA damage, including single-strand breaks, double-strand breaks, and alkali-labile sites. By comparing the extent of DNA damage in treated cells with untreated control cells, the genotoxic potential of a chemical can be determined. The assay can be performed on different cell types, including in vitro cultured cells or even in vivo tissues [117–119].

PDT is a treatment that utilizes non-toxic dyes called photosensitizers (PS) and harmless visible light to generate reactive oxygen species (ROS), by damaging biomolecules and killing cells. While PDT has primarily been developed as a cancer therapy, it can also be used as an antimicrobial approach for localized infections. However, the limited tissue penetration of the longest wavelength used for exciting PS poses a challenge. To address this, Hamblin [120] investigated the use of UCNPs to enable the use of deeper-penetrating near-infrared light for PDT. Specifically, NaYF4 nanoparticles doped with $Yb^{3+}$ and $Er^{3+}$ or $Tm^{3+}$ and $Er^{3+}$ have been attached to PS, allowing NIR light energy to be transduced into ROS, leading to cell killing and tumor regression. Various advancements have been made, including the use of dye-sensitized UCNPs, UCNPs coupled with PS, and the exploration of synergistic drug molecules or techniques. Additionally, the combination of upconversion PDT with different bioimaging modalities has shown promise. However, further research is needed to optimize the drug-delivery capabilities of UCNPs, improve quantum yields, enable intravenous injection and tumor targeting, and ensure safety at required doses before potential clinical applications can be realized.

Nigoghossian et al. [121] focused on evaluating the cytotoxicity, genotoxicity, and mutagenicity of UCNPs incorporated into a 3D-printed composite scaffold with upconversion luminescence for potential biomedical applications. Their system combined the polymer PCL, UCNPs-apatite, and a photosensitizer for PDT in stimulating bone repair. The authors used low concentrations of lanthanide elements in their study and reported no toxicological effects, indicating that the UCNPs incorporated into the scaffold did not induce significant cytotoxicity, genotoxicity, or mutagenicity.

Similarly, Doronkina et al. [122] worked with $NaYF_4$: Er,Yb UCNPs functionalized with $NH_2$-FA and studied erythrocyte aggregation, genotoxicity, and DNA damage. Their findings indicated no genotoxicity or DNA damage associated with the UCNPs. It is important to note that specific results and conclusions may vary depending on the experimental conditions, nanoparticle characteristics, and cell types used in each study. Each study contributes to our understanding of the potential effects and safety profile of UCNPs in various biological systems.

**Table 5** presents the cell viability and surface coating of various UCNP studied by different authors across different types of cells.

Most of the authors reported more than 90% cell viability with functionalized UCNPs, thereby confirming their biocompatibility with the cells under study. These authors utilized several of the techniques mentioned earlier to assess cell viability.

### 4.2 Cellular uptake of nanoparticles

To corroborate the cellular uptake of NPs for in vitro applications, several techniques can be employed to evaluate it. These techniques provide insights into the internalization and localization of NPs within cells. Some commonly used techniques include:

- Flow cytometry: Flow cytometry allows for the quantitative analysis of NPs uptake by cells. It involves labeling NPs with fluorescent dyes or using fluorescently labeled antibodies to detect and quantify the internalized NPs in individual cells. This technique provides information about the percentage of cells that have taken up NPs and can also assess the uptake efficiency.

| Host lattice | Sensitizer/ Activator | Surface coating/ Modification | Cell type | Cell viability | Reference(s) |
|---|---|---|---|---|---|
| $NaYF_4$ | Yb, Er | PEI-FA | BMS cells 1 | ~100%<br>>90% | [123] |
| $NaYF_4$ | Yb, Er | SiO2 - carcinoembryonic antigen (CEA) | HeLa<br>HOS | N/A<br>92.6% | [124] |
| $NaYF_4$ | Yb, Er | SiO2 | BMS cells | 91.10% | [125] |
| $NaYF_4$ | Yb, Er | SiO2 | skeletal myoblast | 93.30% | [126] |
| $NaYF_4$ | Yb, Er | CaF2 shells | HeLa | >90% | [127] |
| $NaYF_4$ | Yb, Er | Ab-siRNA3 | SK-BR-3 cells | 98.60% 92.50% | [128] |
| $NaYF_4$ | Yb, Er | SiO2 | human osteosarcoma | 96.20% | [107] |
| $NaYF_4$ | Yb, Er | No functionalization | HeLa | Not reported | [128] |
| $LaF_3$ | Yb, Ho | mPEG5 | KB cells 4 | >85% | [129] |
| $NaYF_4$ | Yb, Tm | PAA | KB cells | >94% | [103] |
| $Y2O_3$ | Yb, Er | SiO2 -FA | HeLa, MCF7 | >90 | [130] |
| $NaYF_4$ | Yb, Er | HS-PEG<br>HS-DNA | HeLa | Not reported | [131] |
| $NaYF_4$ | Yb, Er | AMPA, APTES, DHCA with POEGA-b-PMAEP (MAEP) | CHO-K1 cells | >95% | [132] |

**Table 5.**
*Surface modification and cell viability of UCNP reported by several authors.*

- Confocal microscopy: Confocal microscopy is a high-resolution imaging technique that allows for the visualization of NPs within cells. By using fluorescent labels or intrinsic luminescence properties of the NPs, confocal microscopy provides spatial information about the distribution and localization of NPs in cellular compartments.

- Epifluorescence microscopy: Epifluorescence microscopy is another imaging technique that utilizes fluorescence to visualize NPs within cells. It provides a wider field of view compared to confocal microscopy but with lower spatial resolution. It can be useful for qualitative assessments of NPs uptake and distribution.

- Fluorometric quantification: This technique involves measuring the fluorescence intensity of NPs, either in cell lysates or supernatants, using a fluorescence plate reader or spectrofluorometer. It allows for the quantitative determination of NPs uptake by cells and can be used to assess the cellular internalization over time.

- Laser scanning microscopy: Laser scanning microscopy, such as laser scanning confocal microscopy or multiphoton microscopy, provides high-resolution imaging capabilities to visualize NPs within cells. It offers the advantage of optical sectioning and 3D imaging, enabling detailed analysis of NPs distribution in different cellular compartments.

The selection of the appropriate technique depends on the specific characteristics of the NPs, the cellular system being studied, and the research objectives. Researchers may choose one or a combination of these techniques to gain a comprehensive understanding of the cellular uptake of NPs and their localization within cells [133].

The laser scanning microscope is a traditional fluorescence technique with the use of fluorescent biomarkers and dyes, but this method lacks the axial (z-axis) resolution to precisely detect the locations of intracellular NPs [134]. This limitation is overcome by employing confocal microscopy, which lends itself to two methods, namely confocal laser scanning microscopy and spinning disk confocal laser microscopy (CLSM). CLSM is widely used in biological research for high-resolution imaging. It employs point illumination and a pinhole aperture to eliminate out-of-focus light, resulting in improved axial resolution. The sample is scanned sequentially by a laser beam, and the emitted fluorescence light is collected through the same objective lens. By using a pinhole to block out-of-focus light, only the in-focus fluorescence from the focal plane is detected, providing sharper images and better localization of NPs within the cell.

Spinning disk confocal microscopy (SDCM) is another variant of confocal microscopy that utilizes a spinning disk with multiple apertures instead of a pinhole. The spinning disk contains an array of microlenses and pinholes, which allows simultaneous illumination and detection of multiple points in the sample. This enables faster image acquisition and is particularly advantageous for capturing rapid cellular processes or live cell imaging. Both CLSM and SDCM provide improved z-axis resolution compared to traditional fluorescence microscopy, enabling the precise detection and localization of intracellular NPs. Researchers can choose between these methods based on their specific experimental requirements, such as imaging speed, sample type, and the need for live cell imaging [135].

The advantages of using UCNPs are the narrow emission spectrum, high expression, and long-term observation in comparison with the dye molecules that presented photo-bleaching, and they are not suitable for long-term observation by using fluorophores. Chatterjee and collaborators [121] reported the first use of NaYF4: Yb, Er UCNPs for cellular imaging functionalized with PEI and conjugated with FA. They worked with human HT29 adenocarcinoma cells and human OVCAR3 ovarian carcinoma cells and confirmed the internalization of the UCNP into both types of cells, the UCNPs emitted in green under a confocal microscope equipped with a 980 nm laser.

Confocal microscopy is an essential imaging technique for visualizing and analyzing the labeled cancer cells, allowing for high-resolution imaging and precise localization of the UCNPs within the cells. It provides researchers with valuable information about the cellular distribution and interaction of UCNPs with specific targets, aiding in the understanding of cancer biology and facilitating the development of targeted therapies.

The study conducted by Chao Wang and collaborators in 2011 [124] focused on the functionalization of UCNPs with polyethylene glycol (PEG) and the subsequent loading of the anticancer drug doxorubicin (DOX) onto the UCNPs. The loading of DOX onto the UCNPs was achieved through simple physical adsorption using a supramolecular chemistry approach. One of the key aspects of this study was the control of DOX loading and release from the UCNPs, which was achieved by manipulating the pH conditions. They observed that the dissociation rate of DOX from the UCNPs increased in an acidic environment, which is favorable for controlled drug release within the intracellular environment. The intracellular delivery of DOX by the functionalized UCNPs was demonstrated through imaging using a confocal microscope. This technique allowed the researchers to visualize and confirm the successful delivery of the drug to the targeted cells. This study highlights the potential of functionalized UCNPs as drug carriers for intracellular drug delivery.

The ability to control drug release in response to specific conditions, such as pH, offers opportunities for targeted and controlled drug delivery systems with reduced side effects. The use of confocal microscopy for imaging provides valuable insights into the cellular uptake and intracellular distribution of the drug-loaded UCNPs [5].

For *in vivo* applications several authors reported internalization and visualization of UCNP with different techniques. The limit for cell detection via UCNPs imaging is a relevant parameter to determine the sensitivity of the in vivo imaging technique by UCNPs. Liu and coworkers [120] proved that the 800 nm emissions from UCNPs consisting of sub-10 nm hexagonal NaLuF4: Yb can penetrate through a whole black mouse with a depth of ~2 cm, under excitation with a 980 nm laser using a confocal microscope. Also, they reported the study of lymphatic drainage in various parts of the body, for the diagnosis and treatment of cancer. They developed a UCNP of LaF3: Yb, Tm and applied it for lymphatic system imaging in vivo with a high signal-to-noise ratio (~33) of the UCNP signal (em = 800 nm). An intense UCNP signal was observed from the removed lymph node, whereas no obvious UCNP signal was detected from the area outside of the draining lymphatic system.

Epifluorescence microscopy is a useful technique for imaging thick samples, particularly those that are over 10 μm deep. However, it has some limitations, such as the potential for high background signal due to intense illumination and excitation of molecules outside the focal plane. Despite this, several authors have successfully utilized epifluorescence microscopy for bioimaging with UCNPs.

One study by Mrazek and colleagues [107] introduced a modification to the epifluorescence microscope that enabled quantitative widefield imaging of UCNPs. They incorporated a tandem microlens array, a moving diffuser, and a telescope into their setup. By adjusting the top-hat area to the field of view, they achieved a high level of illumination profile homogeneity, reaching approximately 98%. This modified illuminator was combined with the epifluorescence attachment of the microscope, allowing for easy switching between observation of UCNPs and traditional fluorescent dyes.

In another study by Yang and colleagues [129], a quantitative fluorescence image analysis based on multicolor UCNPs was developed for the detection of ochratoxin A and zearalenone. They first obtained a clear image of the UCNPs using near-infrared (NIR) excitation. Then, with the appropriate filters and detection methods, they were able to detect the presence of the target analytes using the fluorescence signals emitted by the UCNPs. These studies demonstrate the successful utilization of epifluorescence microscopy for imaging UCNPs and their applications in quantitative analysis. The modifications made to the microscopy setup and the careful selection of excitation and detection parameters contribute to enhancing the sensitivity and accuracy of the imaging process.

Flow cytometry is indeed a powerful technique for the analysis of cellular internalization of fluorescently labeled NPs. It allows for the quantitative measurement of the proportion of cells that have taken up the labeled NPs. Flow cytometry utilizes laser-based excitation and detection of fluorescence signals emitted by the labeled NPs, providing a high statistical analysis of a large number of cells. One of the main advantages of flow cytometry is its ability to analyze a large population of cells rapidly, providing statistical data on the percentage of cells that have internalized the NPs. It can also provide information on the level of NP uptake within individual cells by measuring the fluorescence intensity.

However, it is important to note that flow cytometry has its limitations. One of the main drawbacks is its relatively low resolution, which makes it challenging to study the precise localization of NPs within different cellular compartments. Flow cytometry primarily provides information on whether the NPs have been internalized by the cells, but it cannot determine their specific subcellular localization or membrane-bound interactions. Therefore, while flow cytometry is an excellent method for the detection and quantification of cellular internalization of NPs, it is often complemented by other techniques such as fluorescence microscopy or electron microscopy to gain more detailed information about NP localization within cells and their interactions with cellular structures.

Juarez-Moreno et al. analyzed the internalization of UCNPs functionalized with FA in B16-F10 melanoma cancer cells, and for negative control they used the L929 cell line because it does not have folic acid receptors on the membrane. The uptake of FA-UCNPs into cells was determined in terms of cell granularity, also known as intracellular complexity, which was detected by the side scatter beam (SSC-H) of the flow cytometer. They compared the internalization of the UCNPs into 2D and 3D cell arrays and concluded that the internalization of the nanoparticles was concentration dependent. The internalization of FA-UCNPs was higher in melanoma B16-F10 cells in a 3D conformation than in 2D cultures [130].

Indeed, the unique properties of UCNPs make them highly attractive for bioimaging applications. UCNPs offer several advantages over conventional biomarkers and therapeutic agents, making them promising candidates for various biomedical applications. However, further research is still needed to fully understand the behavior of UCNPs in complex biological systems, especially in live animals. This includes studying their biodistribution, pharmacokinetics, long-term toxicity, and clearance mechanisms. Comprehensive nanotoxicological assessments are crucial to ensure the safe use of UCNPs in biomedical applications.

By gaining a deeper understanding of the behavior and potential risks associated with UCNPs, researchers can develop strategies to optimize their properties, improve their biocompatibility, and minimize any potential adverse effects. This knowledge will contribute to the development of safe and effective UCNPs for various biomedicine applications, including bioimaging, drug delivery, and theranostic.

Overall, UCNPs hold great promise in the field of biomedicine, but continued research and evaluation are necessary to unlock their full potential and ensure their safe and efficient use in clinical settings.

## 5. Conclusions

In summary, upconversion nanoparticles (UCNPs) exhibit great promise in the realms of bioimaging and theranostic applications, thanks to their efficient near-infrared-triggered emissions, large anti-Stokes shifts, high sensitivity, and biocompatibility. Despite their potential, several challenges necessitate attention for optimal utilization:

1. Emission Properties: The surface properties of UCNPs can impact emission ratios, leading to monochromatic emissions. Addressing this involves refining the synthesis process by selecting appropriate host lattices and activators.

2. Quantum Yield and Deep Tissue Imaging: Low quantum yields of $Ln^{3+}$ ions limit UCNPs' applications for deep tissue imaging. Strategies involve enhancing

upconversion efficiency through doping UCNPs and exploring alternative activators like $Yb^{3+}$.

3. Expansion to Other Biological Models: While promising *in vitro* and small animal studies, UCNPs' application in diverse biological models, such as plants or large animals, is limited and requires further testing.

4. Tumor Targeting and Intravenous Injection: Improving targeting capabilities to specific tumor tissues and enabling intravenous injection would enhance precision in cancer therapy.

5. Drug Delivery and Therapeutics: Optimizing UCNPs as carriers for therapeutic molecules requires focusing on controlled release, stimuli-responsive behavior, and multi-functional payloads for targeted therapies.

6. Imaging and Sensing Modalities: Integrating UCNPs with various bioimaging techniques and sensing platforms could improve diagnostic capabilities and open new avenues for biosensing.

7. Scalability and Cost-effectiveness: Streamlining synthesis methods for scalable and cost-effective manufacturing is crucial for widespread adoption.

8. Multifunctional Platforms: Exploring integration with other nanomaterials or technologies could lead to the development of multifunctional platforms with enhanced properties.

9. Overheating Issue: Exploring alternative excitation wavelengths, such as 880 nm NIR, and optimizing activator choices, especially $Yb^{3+}$, can mitigate the overheating issue associated with the typical 980 nm laser.

10. Nanotoxicological Assessment: Further analysis and a comprehensive approach considering LNPs architecture, cell biology, and predictive computer-aided tools are necessary for a deeper understanding of nanotoxicity.

Overall, the field of UCNPs for bioimaging and chemotherapeutic delivery holds enormous potential. By addressing the challenges mentioned and conducting further research, the development of next-generation nano-engineered LNPs with enhanced properties and translational potential for clinical applications is within reach.

## Acknowledgements

The authors wish to acknowledge the support of CETYS University and Universidad Latina de Panama.

## Conflicts of interest

The authors declare no conflict of interest.

## Author details

Dalia Chávez-García[1*] and Mario Guzman[2]

1 Universidad Latina de Panama, Panama

2 Centro de Enseñanza Técnica y Superior (CETYS), Ensenada, México

*Address all correspondence to: daliachavez@ulatina.edu.pa

## References


[1] Rihn BH. Biomedical Application of Nanoparticles. 1st ed. Boca Raton, FL: Taylor & Francis Group L, CRC Press; 2018. 321 p

[2] Zhong Y, Dai H. A mini-review on rare-earth down-conversion nanoparticles for NIR-II imaging of biological systems. In: Nano Research. Vol. 13, no. 5. Beijing, China: Tsinghua University Press; 2020. pp. 1281-1294. DOI: 10.1007/s12274-020-2721-0

[3] Wan J, Wang Z, Chen X, Mu L, Qian Y. Shape-tailored photoluminescent intensity of red phosphor $Y_2O_3$: $Eu^{3+}$. Journal of Crystal Growth. 2005;**284**:538-543

[4] Wang W, Sun Z, He X, Wei Y, Zou Z, Zhang J. How to design ultraviolet emitting persistent materials for potential multifunctional applications: A living example of a $NaLuGeO_4$:$Bi^{3+}$, $Eu^{3+}$ phosphor. Journal of Materials Chemistry C Materials. 2017;**5**(17):4310-4318. DOI: 10.1039/C6TC05598B

[5] Wang M, Abbineni G, Clevenger A, Mao C, Xu S. Upconversion nanoparticles: Synthesis, surface modification and biological applications. Nanomedicine: Nanotechnology, Biology, and Medicine. 2011;7(6):710-729. DOI: 10.1016/j.nano.2011.02.013

[6] Blasse G, Grabmaier BC. Luminescent Materials. Berlin Heidelberg: Springer-Verlag; 1994. p. 230. DOI: 10.1021/ja965667t

[7] Blasse G, Grabmaier BC. A general introduction to luminescent materials. In: Luminescent Materials. Berlin, Heidelberg: Springer; 2011. DOI: 10.1007/978-3-642-79017-1_1

[8] Blasse G, Grabmaier BC. How does a luminescent material absorb its excitation energy? In: Luminescent Materials. Berlin, Heidelberg: Springer Science & Business Media; 2011. DOI: 10.1007/978-3-642-79017-1_2

[9] Vetrone F, Boyer JC, Capobianco JA, Speghini A, Bettinelli M. Significance of $Yb^{3+}$ concentration on the upconversion mechanisms in codoped $Y_2O_3$: $Er^{3+}$, $Yb^{3+}$ nanocrystals. Journal of Applied Physics. 2004;**96**(1):661-667. DOI: 10.1063/1.1739523

[10] Chavez D, Contreras HG. Synthesis and upconversion luminescence of nanoparticles $Y_2O_3$ and $Gd_2O_3$ Co-doped with $Yb^{3+}$ and $Er^{3+}$. Nanomaterials and Nanotechnology. 2016;**6**:7. DOI: 10.5772/62188

[11] Lojpur V, Nikolic M, Mancic L, Milosevic O, Dramicanin MD. $Y_2O_3$: Yb, Tm and $Y_2O_3$:Yb, Ho powders for low-temperature thermometry based on up-conversion fluorescence. Ceramics International. 2013;**39**(2):1129-1134

[12] Su J, Song F, Tan H, Han L, Zhou F, Tian J. Phonon-assisted mechanisms and concentration dependence of $Tm^{3+}$ blue upconversion luminescence in codoped $NaY(WO_4)_2$ crystals. Journal of Physics D: Applied Physics. 2006;**39**(10):2094-2099. DOI: 10.1088/0022-3727/39/10/019

[13] Šević D, Rabasović MS, Križan J, Savić-Šević S, Rabasović MD, Marinkovic BP. Effects of temperature on luminescent properties of $Gd_2O_3$: Er, Yb nanophosphor. Optical and Quantum Electronics. 2020;**52**(5):232. DOI: 10.1007/s11082-020-02348-y

[14] Yang J, Zhang C, Peng C, Li C, Wang L, Chai R. Controllable red, green, blue (RGB) and bright white upconversion luminescence of

$Lu_2O_3$: $Yb^{3+}/Er^{3+}/Tm^{3+}$ nanocrystals through single laser excitation at 980 nm. Chemistry – A European Journal [Internet]. 2009;**15**(18):4649-4655. DOI: 10.1002/chem.200802106

[15] Han Q, Yan T, Song Y, Wang Y, Zhang X. Up-conversion luminescence enhancement and tunable emission color of $Y_2O_3$: Tm/Yb@$Y_2O_3$: Yb@$Y_2O_3$: Er/Yb multilayer thin films. Journal of Alloys and Compounds. 2021;**877**:160151. DOI: 10.1016/j.jallcom.2021.160151

[16] Das GK, Tan TTY. Rare-earth-doped and codoped $Y_2O_3$ nanomaterials as potential bioimaging probes. Journal of Physical Chemistry C. 2008;**112**(30):11211-11217. DOI: 10.1021/jp802076n

[17] Chávez-García D, Juárez-Moreno K, Hirata GA. Upconversion nanoparticles $Y_2O_3$ and $Gd_2O_3$ Co-doped with $Er^{3+}$ and $Yb^{3+}$ with aminosilane-folic acid functionalization for breast and cervix cancer cells detection. MRS Advances. 2017;**2**(52):2983-2988. DOI: 10.1557/adv.2017.447

[18] Hemmer E, Venkatachalam N, Hyodo H, Hattori A, Ebina Y, Kishimoto H. Upconverting and NIR emitting rare earth based nanostructures for NIR-bioimaging. Nanoscale. 2013;**5**(23):11339-11361. DOI: 10.1039/C3NR02286B

[19] Liu C, Chen D. Controlled synthesis of hexagon shaped lanthanide-doped LaF3 nanoplates with multicolor upconversion fluorescence. Journal of Materials Chemistry. 2007;**17**(37):3875-3880. DOI: 10.1039/B707927C

[20] Pires AM, Serra OA, Davolos MR. Yttrium oxysulfide nanosized spherical particles doped with Yb and Er or Yb and Tm: Efficient materials for up-converting phosphor technology field. Journal of Alloys and Compounds. 2004;**374**:181-184. DOI: 10.1016/j.jallcom.2003.11.088

[21] Steblevskaya NI, Medkov MA, Belobeletskaya MV. Luminophores and protective coatings based on oxides and oxysulfides of rare-earth elements prepared by extraction pyrolysis. Theoretical Foundations of Chemical Engineering. 2014;**48**(4):449-453. DOI: 10.1134/S0040579514040125

[22] De HP, Huang XG, Zhang QT. Effect of doping concentration on the up-conversion luminescent properties of $Y_2O_3$:Yb, Tm phosphors. Advanced Materials Research. 2011;**197-198**:558-562. DOI: 10.4028/www.scientific.net/AMR.197-198.558

[23] Hirai T, Orikoshi T. Preparation of $Gd_2O_3$: Yb, Er and $Gd_2O_2S$: Yb, Er infrared-to-visible conversion phosphor ultrafine particles using an emulsion liquid membrane system. Journal of Colloid and Interface Science. 2004;**269**(1):103-108. DOI: 10.1016/j.jcis.2003.08.026

[24] Kataria V, Mehta DS. Investigation of concurrent emissions in visible, UV and NIR region in $Gd_2O_2S$: Er, Yb nanophosphor by diverse excitation wavelengths as a function of firing temperature. Optical Materials (Amst). 2019;**95**:109204. DOI: 10.1016/j.optmat.2019.109204

[25] Du YP, Zhang YW, Sun LD, Yan CH. Luminescent monodisperse nanocrystals of lanthanide oxyfluorides synthesized from trifluoroacetate precursors in high-boiling solvents. The Journal of Physical Chemistry C. 2008;**112**(2):405-415. DOI: 10.1021/jp076717r

[26] Gao D, Gao F, Wu J, Kuang Q, Xing C, Chen W. Up-conversion luminescence performance and application of GdOF: Yb, Er porous spheres obtained by calcining $NaGdF_4$:

Yb, Er microcrystals. Applied Surface Science. 2022;**587**:152820. DOI: 10.1016/j.apsusc.2022.152820

[27] Gao L, Ge X, Chai Z, Xu G, Wang X, Wang C. Shape-controlled synthesis of octahedral α-$NaYF_4$ and its rare earth doped submicrometer particles in acetic acid. Nano Research. 2009;**2**(7):565-574. DOI: 10.1007/s12274-009-9056-1

[28] Kang M, Kang HB, Park S, Jang HS. Facile synthesis of sub-10 nm-sized bright red-emitting upconversion nanophosphors via tetrahedral YOF: Yb, Er seed-mediated growth. Chemical Communications. 2019;**55**(89):13350-13353. DOI: 10.1039/C9CC06797C

[29] AitMellal O, Oufni L, Messous MY, Trandafir MM, Chirica IM, Florea M. Comparative investigation of structural, EPR, optical and photoluminescence properties of nanostructured $LaPO_4$: Ce/RE/me and $LaPO_4$: Yb/Er phosphors prepared by co-precipitation method. Journal of Solid State Chemistry. 2021;**301**:122310. DOI: 10.1016/j.jssc.2021.122310

[30] Lisiecki R, Ryba-Romanowski W, Speghini A, Bettinelli M. Luminescence spectroscopy of $Er^{3+}$–doped and $Er^{3+}$, $Yb^{3+}$–codoped $LaPO_4$ single crystals. Journal of Luminescence. 2009;**129**(5):521-525. DOI: 10.1016/j.jlumin.2008.12.006

[31] Heer S, Lehmann O, Haase M, Güdel HU. Blue, green, and red upconversion emission from lanthanide-doped $LuPO_4$ and $YbPO_4$ nanocrystals in a transparent colloidal solution. Angewandte Chemie International Edition. 2003;**42**(27):3179-3182. DOI: 10.1002/anie.200351091

[32] Yi G, Sun B, Yang F, Chen D, Zhou Y, Cheng J. Synthesis and characterization of high-efficiency nanocrystal up-conversion phosphors: Ytterbium and erbium codoped lanthanum molybdate. Chemistry of Materials. 2002;**14**(7):2910-2914. DOI: 10.1021/cm0115416

[33] Chen Z, Bu W, Zhang N, Shi J. Controlled construction of monodisperse $La_2(MoO_4)_3$: Yb, Tm microarchitectures with upconversion luminescent property. The Journal of Physical Chemistry C. 2008;**112**(11):4378-4383. DOI: 10.1021/jp711213r

[34] Zhou Y, He XH, Yan B. Self-assembled RE2$(MO_4)_3$: $Ln^{3+}$ (RE=Y, La, Gd, Lu; M=W, Mo; ln=Yb/Er, Yb/Tm) hierarchical microcrystals: Hydrothermal synthesis and up-conversion luminescence. Optical Materials (Amst). 2014;**36**(3):602-607. DOI: 10.1016/j.optmat.2013.10.036

[35] Yu W, Tian Y, Xing M, Fu Y, Zhang H, Luo X. Up-conversion luminescence of $NaY(WO_4)_2$: Yb, Er under 1550 and 980 nm excitation. Materials Research Bulletin. 2016;**80**:223-229. DOI: 10.1016/j.materresbull.2016.03.013

[36] Xue N, Fan X, Wang Z, Wang M. Synthesis process and luminescence properties of $Ln^{3+}$ doped $NaY(WO_4)_2$ nanoparticles. Materials Letters. 2007;**61**(7):1576-1579. DOI: 10.1016/j.matlet.2006.07.082

[37] Sun C, Lü W, Yang F, Tu C. Tunable red-green-blue multicolor luminescence in $Yb^{3+}/Tm^{3+}/Ho^{3+}$: Gd3Ga5$O_{12}$ nano-crystals. Journal of Alloys and Compounds. 2012;**512**(1):160-164. DOI: 10.1016/j.jallcom.2011.09.056

[38] Pandozzi F, Vetrone F, Boyer JC, Naccache R, Capobianco JA, Speghini A. A spectroscopic analysis of blue and ultraviolet upconverted emissions from $Gd_3Ga5O_{12}$: $Tm^{3+}$, $Yb^{3+}$ nanocrystals. Journal of Physical Chemistry B. 2005;**109**(37):17400-17405. DOI: 10.1021/jp052192w

[39] Tsuboi T. Upconversion emission in $Er^{3+}/Yb^{3+}$–codoped $YVO_4$ crystals. Physical Review B. 2020;**62**(7):4200. DOI: 10.1103/PhysRevB.62.4200

[40] Ma R, Shimmon R, McDonagh A, Maynard P, Lennard C, Roux C. Fingermark detection on non-porous and semi-porous surfaces using $YVO_4$: Er, Yb luminescent upconverting particles. Forensic Science International. 2012;**217**(1):e23-e26. DOI: 10.1016/j.forsciint.2011.10.033

[41] Wang K, Ma J, He M, Gao G, Xu H, Sang J, et al. Toxicity assessments of near-infrared upconversion luminescent LaF3: Yb, Er in early development of zebrafish embryos. Theranostics. 2013;**3**(4):258-266. DOI: 10.7150/thno.5701

[42] Yi GS, Chow GM. Colloidal LaF3: Yb, Er, LaF3: Yb, Ho and LaF3: Yb, Tm nanocrystals with multicolor upconversion fluorescence. Journal of Materials Chemistry. 2005;**15**(41):4460-4464. DOI: 10.1039/B508240D

[43] Huang X. Enhancement of near-infrared to near-infrared upconversion luminescence in sub-10-nm ultra-small LaF3:Yb3+/Tm3+ nanoparticles through lanthanide doping. Optics Letters. 2015;**40**(22):5231-5234. Available from: https://opg.optica.org/ol/abstract.cfm?URI=ol-40-22-5231

[44] Wang X, Hu Z, Sun M, Du P, Liu W, Huang S, et al. Phase-conversion synthesis of LaF3: Yb/RE (RE = Ho, Er) nanocrystals with $Ln_2(OH)_4SO_4·2H_2O$ type layered compound as a new template, phase/morphology evolution, and upconversion luminescence. Journal of Materials Research and Technology. 2020;**9**(5):10659-10668. Available from: https://www.sciencedirect.com/science/article/pii/S2238785420315684

[45] Hu H, Chen Z, Cao T, Zhang Q, Yu M, Li F, et al. Hydrothermal synthesis of hexagonal lanthanide-doped LaF3 nanoplates with bright upconversion luminescence. Nanotechnology. 2008;**19**(37):375702. DOI: 10.1088/0957-4484/19/37/375702

[46] Xie J, Hu W, Tian D, Wei Y, Zheng G, Huang L, et al. Selective growth and upconversion photoluminescence of Y-based fluorides: From $NaYF_4$: Yb/Er to YF3: Yb/Er crystals. Nanotechnology [Internet]. 2020;**31**(50):505605. DOI: 10.1088/1361-6528/abb627

[47] Suo H, Hu F, Zhao X, Zhang Z, Li T, Duan C, et al. All-in-one thermometer-heater up-converting platform YF3: $Yb^{3+}$, $Tm^{3+}$ operating in the first biological window. Journal of Materials Chemistry C Materials [Internet]. 2017;**5**(6):1501-1507. DOI: 10.1039/C6TC05449H

[48] Yan RX, Li YD. Down/up conversion in $Ln^{3+}$-doped YF3 nanocrystals. Advanced Functional Materials [Internet]. 2005;**15**(5):763-770. DOI: 10.1002/adfm.200305044

[49] Bartkowiak A, Siejca A, Borkowski K, Lis S, Grzyb T. Up-converting LuF3 and NaLuF4 fluorides doped with $Yb^{3+}/Er^{3+}$ or $Yb^{3+}/Tm^{3+}$ ions for latent fingermarks detection. Journal of Alloys and Compounds. 2019;**784**:641-652. DOI: 10.1016/j.jallcom.2018.12.344

[50] Wang G, Qin W, Zhang J, Zhang J, Wangyan CC, et al. Synthesis, growth mechanism, and tunable upconversion luminescence of $Yb^{3+}/Tm^{3+}$-codoped YF3 nanobundles. The Journal of Physical Chemistry C [Internet]. 2008;**112**(32):12161-12167. DOI: 10.1021/jp8004713

[51] Homann C, Krukewitt L, Frenzel F, Grauel B, Würth C, Resch-Genger U, et al. NaYF4: Yb,Er/NaYF4 core/shell nanocrystals with high upconversion luminescence quantum yield.

Angewandte Chemie International Edition. 2018;**57**(28):8765-8769. DOI: 10.1002/anie.201803083

[52] Yuan Z, Lin C, Wang Y, Niu Y, Yan J, Lu M, et al. Anion additive-induced size, morphology, and local structure tuning of lanthanide-doped upconversion nanoparticles. Advanced Materials Interfaces [Internet]. 2022;**9**:2201277. DOI: 10.1002/admi.202201277

[53] Wang M, Liu JL, Zhang YX, Hou W, Wu XL, Xu SK. Two-phase solvothermal synthesis of rare-earth doped NaYF4 upconversion fluorescent nanocrystals. Materials Letters. 2009;**63**(2):325-327. DOI: 10.1016/j.matlet.2008.10.028

[54] Tang Y, Di W, Zhai X, Yang R, Qin W. NIR-responsive photocatalytic activity and mechanism of NaYF4: Yb, Tm@$TiO_2$ Core–Shell nanoparticles. ACS Catalysis [Internet]. 2013;**3**(3):405-412. DOI: 10.1021/cs300808r

[55] Lingeshwar Reddy K, Srinivas V, Shankar KR, Kumar S, Sharma V, Kumar A, et al. Enhancement of luminescence intensity in red emitting $NaYF_4$: Yb/Ho/Mn upconversion nanophosphors by variation of reaction parameters. The Journal of Physical Chemistry C [Internet]. 2017;**121**(21):11783-11793. DOI: 10.1021/acs.jpcc.7b01334

[56] Hong AR, Kim SY, Cho SH, Lee K, Jang HS. Facile synthesis of multicolor tunable ultrasmall LiYF4: Yb, Tm, Er/LiGdF4 core/shell upconversion nanophosphors with sub-10 nm size. Dyes and Pigments. 2017;**139**:831-838. DOI: 10.1016/j.dyepig.2016.12.048

[57] Naccache R, Vetrone F, Mahalingam V, Cuccia LA, Capobianco JA. Controlled synthesis and water dispersibility of hexagonal phase $NaGdF_4$: $Ho^{3+}/Yb^{3+}$ nanoparticles. Chemistry of Materials [Internet]. 2009;**21**(4):717-723. DOI: 10.1021/cm803151y

[58] Liang L, Zhang X, Hu H, Feng L, Wu M, Su Q. Up-conversion properties in KGd2F7: $Yb^{3+}/Er^{3+}$. Materials Letters. 2005;**59**(17):2186-2190. DOI: 10.1016/j.matlet.2005.02.061

[59] Yu L, Li G, Liu Y, Jiang F, Hong M. Lanthanide-doped KGd2F7 nanocrystals: Controlled synthesis, optical properties, and spectroscopic identification of the optimum core/shell architecture for highly enhanced upconverting luminescence. Crystal Growth & Design. 2019;**19**(4):2340-2349. DOI: 10.1021/acs.cgd.9b00040

[60] Niu N, Yang P, Liu Y, Li C, Wang D, Gai S, et al. Controllable synthesis and up-conversion properties of tetragonal BaYF5:Yb/ln (ln=Er, Tm, and Ho) nanocrystals. Journal of Colloid and Interface Science. 2011;**362**(2):389-396. DOI: 10.1016/j.jcis.2011.07.001

[61] Luo R, Chen L, Li Q, Zhou J, Mei L, Ning Z, et al. Bi3+-doped BaYF5:Yb,Er Upconversion nanoparticles with enhanced luminescence and application case for X-ray computed tomography imaging. Inorganic Chemistry. 2020;**59**(24):17906-17915. DOI: 10.1021/acs.inorgchem.0c01818

[62] Vetrone F, Mahalingam V, Capobianco JA. Near-infrared-to-blue upconversion in colloidal BaYF5: $Tm^{3+}$, $Yb^{3+}$ nanocrystals. Chemistry of Materials. 2009;**21**(9):1847-1851. DOI: 10.1021/cm900313s

[63] Smet PF, Moreels I, Hens Z, Poelman D. Luminescence in sulfides: A rich history and a bright future. Materials [Internet]. 2010;**3**(4):2834-2883. DOI: 10.3390/ma3042834

[64] Wu W, He Q, Jiang C. Magnetic iron oxide nanoparticles: Synthesis and surface functionalization strategies. Nanoscale Research Letters. 2008;**3**(11):397-415. DOI: 10.1007/s11671-008-9174-9

[65] Priya R, Pandey OP, Dhoble SJ. Review on the synthesis, structural and photo-physical properties of $Gd_2O_3$ phosphors for various luminescent applications. Optics and Laser Technology. 2021;**135**:106663. DOI: 10.1016/j.optlastec.2020.106663

[66] Fang Z, Yu H, Zhang B, Jiang D, Wu Q, Xiong L, et al. Suppression of $Eu^{2+}$ luminescence and enhancement of $Eu^{3+}$ emission in Eu: $CaF_2$ single crystal via $Gd^{3+}$ co-doping. Journal of Luminescence. 2021;**233**:117877. DOI: 10.1016/j.jlumin.2020.117877

[67] Werts MHV. Making sense of lanthanide luminescence. Science Progress. 2005;**88**(2):101-131. DOI: 10.3184/003685005783238435

[68] Iqbal S, Fakhar-e-Alam M, Alimgeer KS, Atif M, Hanif A, Yaqub N, et al. Mathematical modeling and experimental analysis of the efficacy of photodynamic therapy in conjunction with photo thermal therapy and PEG-coated Au-doped $TiO_2$ nanostructures to target MCF-7 cancerous cells. Saudi Journal of Biological Sciences. 2021;**28**(2):1226-1232. DOI: 10.1016/j.sjbs.2020.11.086

[69] Dos Santos CCL, Passos Farias IA, de Jesus dos Reis Albuquerque A, Silva DPMFE, da Costa One GM, Sampaio FC. Antimicrobial activity of nano cerium oxide (IV) ($CeO_2$) against Streptococcus mutans. BMC Proceedings. 2014;**8**(4):P48. DOI: 10.1186/1753-6561-8-S4-P48

[70] Barar J, Kafil V, Majd MH, Barzegari A, Khani S, Johari-Ahar M, et al. Multifunctional mitoxantrone-conjugated magnetic nanosystem for targeted therapy of folate receptor-overexpressing malignant cells. Journal of Nanobiotechnology. 2015;**13**(1):26. DOI: 10.1186/s12951-015-0083-7

[71] Park J, Joo J, Kwon SG, Jang Y, Hyeon T. Synthesis of monodisperse spherical nanocrystals. Angewandte Chemie International Edition. 2007;**46**(25):4630-4660. DOI: 10.1002/anie.200603148

[72] Taxak VB, Khatkar SP, Do HS, Kumar R, Kumar M. Tartaric acid-assisted sol – gel synthesis of $Y_2O_3$: $Eu^{3+}$ nanoparticles. Journal of Alloys and Compounds. 2009;**469**:224-228. DOI: 10.1016/j.jallcom.2008.01.088

[73] Talane TE, Mbule PS, Noto LL, Shingange K, Mhlongo GH, Mothudi BM, et al. Sol-gel preparation and characterization of $Er^{3+}$ doped $TiO_2$ luminescent nanoparticles. Materials Research Bulletin. 2018;**108**:234-241. DOI: 10.1016/j.materresbull.2018.09.007

[74] Ai X, Aw J, Xing B. Upconversion nanoparticles for bioimaging. In: Liu RS, editor. Phosphors, Up Conversion Nano Particles, Quantum Dots and their Applications. Vol. 2. Singapore: Springer Singapore; 2016. pp. 363-390. DOI: 10.1007/978-981-10-1590-8_12

[75] Boyer JC, Vetrone F, Cuccia LA, Capobianco JA. Synthesis of colloidal upconverting NaYF4 nanocrystals doped with $Er^{3+}$, $Yb^{3+}$ and $Tm^{3+}$, $Yb^{3+}$ via thermal decomposition of lanthanide trifluoroacetate precursors. Journal of the American Chemical Society. 2006;**128**(23):7444-7445. DOI: 10.1021/ja061848b

[76] Li C, Yang J, Quan Z, Yang P, Kong D, Lin J. Different microstructures of β-NaYF4 fabricated by hydrothermal

process: Effects of pH values and fluoride sources. Chemistry of Materials. 2007;**19**(20):4933-4942. DOI: 10.1021/cm071668g

[77] Yi G, Lu H, Zhao S, Ge Y, Yang W, Chen D, et al. Synthesis, characterization, and biological application of size-controlled nanocrystalline $NaYF_4$: Yb, Er infrared-to-visible up-conversion phosphors. Nano Lett. 2004;**4**(11):2191-2196. DOI: 10.1021/nl048680h

[78] Pandey A, Rai VK, Dey R, Kumar K. Enriched green upconversion emission in combustion synthesized $Y_2O_3$: $Ho^{3+}$–$Yb^{3+}$ phosphor. Materials Chemistry and Physics. 2013;**139**(2):483-488. Available from: https://www.sciencedirect.com/science/article/pii/S0254058413000874

[79] Li Y, Chen R, Li Y, Sharafudeen K, Liu S, Wu D, et al. Folic acid-conjugated chromium(III) doped nanoparticles consisting of mixed oxides of zinc, gallium and tin, and possessing near-infrared and long persistent phosphorescence for targeted imaging of cancer cells. Microchimica Acta. 2015;**182**(9-10):1827-1834. DOI: 10.1007/s00604-015-1486-8

[80] Sudimack J, Lee RJ. Targeted drug delivery via the folate receptor. Advanced Drug Delivery Reviews. 2000;**41**(2):147-162. DOI: 10.1016/S0169-409X(99)00062-9

[81] Chakraborti S, Chakraborty S, Saha S, Manna A, Banerjee S, Adhikary A, et al. PEG-functionalized zinc oxide nanoparticles induce apoptosis in breast cancer cells through reactive oxygen species-dependent impairment of DNA damage repair enzyme NEIL2. Free Radical Biology & Medicine. 2017;**103**:35-47. DOI: 10.1016/j.freeradbiomed.2016.11.048

[82] Cheng J, Teply BA, Sherifi I, Sung J, Luther G, Gu FX, et al. Formulation of functionalized PLGA-PEG nanoparticles for *in vivo* targeted drug delivery. Biomaterials. 2007;**28**(5):869-876. DOI: 10.1016/j.biomaterials.2006.09.047

[83] Manson J, Kumar D, Meenan BJ, Dixon D. Polyethylene glycol functionalized gold nanoparticles: The influence of capping density on stability in various media. Gold Bulletin. 2011;**44**(2):99-105. DOI: 10.1007/s13404-011-0015-8

[84] Maldiney T, Richard C, Seguin J, Wattier N, Bessodes M, Scherman D. Effect of core diameter, surface coating, and PEG chain length on the biodistribution of persistent luminescence nanoparticles in mice. ACS Nano. 2011;**5**(2):854-862. DOI: 10.1021/nn101937h

[85] Buchman YK, Lellouche E, Zigdon S, Bechor M, Michaeli S, Lellouche JP. Silica nanoparticles and polyethyleneimine (PEI)-mediated functionalization: A new method of PEI covalent attachment for siRNA delivery applications. Bioconjugate Chemistry. 2013;**24**(12):2076-2087. DOI: 10.1021/bc4004316

[86] Liu C, Wang Z, Wang X, Li Z. Surface modification of hydrophobic NaYF4: Yb, Er upconversion nanophosphors and their applications for immunoassay. Science China. Chemistry. 2011;**54**(8):1292-1297. DOI: 10.1007/s11426-011-4319-6

[87] Rao J, Dragulescu-Andrasi A, Yao H. Fluorescence imaging *in vivo*: Recent advances. Current Opinion in Biotechnology [Internet]. 2007;**18**(1):17-25. DOI: 10.1016/j.copbio.2007.01.003

[88] Yang D, Dai Y, Ma P, Kang X, Cheng Z, Li C, et al. One-step synthesis of small-sized and water-soluble NaREF4 upconversion nanoparticles for In vitro cell

imaging and drug delivery. Chemistry – A European Journal. 2013;**19**(8):2685-2694. DOI: 10.1002/chem.201203634

[89] Suvorov NV, Mironov AF, Grin MA. Folic acid and its derivatives for targeted photodynamic therapy of cancer. Russian Chemical Bulletin. 2017;**66**:1982-2008. DOI: 10.1007/s11172-017-1973-7

[90] Xie A, Tsvetkova I, Liu Y, Ye X, Hewavitharanage P, Dragnea B, et al. Hydrophobic cargo encapsulation into virus protein cages by self-assembly in an aprotic organic solvent. Bioconjugate Chemistry. 2021;**32**(11):2366-2376. DOI: 10.1021/acs.bioconjchem.1c00420

[91] Klymchenko A, Duportail G, Mely Y. Fluorescent methods to study biological membranes. Springer Series on Fluorescence. 2013;**13**:51-70. DOI: 10.1007/4243_2012_44

[92] Hong G, Antaris AL, Dai H. Near-infrared fluorophores for biomedical imaging. Nature Biomedical Engineering [Internet]. 2017;**1**(1):10. DOI: 10.1038/s41551-016-0010

[93] Hsu HJ, Bugno J, Sr L, Hong S. Dendrimer-based nanocarriers: A versatile platform for drug delivery. Wiley Interdisciplinary Reviews - Nanomedicine and Nanobiotechnology. 2017;**9**:e1409. DOI: 10.1002/wnan.1409

[94] Dong H, Du SR, Zheng XY, Lyu GM, Sun LD, Li LD, et al. Lanthanide nanoparticles: From design toward bioimaging and therapy. Chemical Reviews. American Chemical Society. 2015;**115**:10725-10815. DOI: 10.1021/acs.chemrev.5b00091

[95] Yu M, Li F, Chen Z, Hu H, Zhan C, Yang H, et al. Laser scanning up-conversion luminescence microscopy for imaging cells labeled with rare-earth nanophosphors. Analytical Chemistry. 2009;**81**(3):930-935. DOI: 10.1021/ac802072d

[96] Jin J, Gu YJ, Man CWY, Cheng J, Xu Z, Zhang Y, et al. Polymer-coated NaYF4: $Yb^{3+}$, $Er^{3+}$ upconversion nanoparticles for charge-dependent cellular imaging. ACS Nano. 2011;**5**(10):7838-7847. DOI: 10.1021/nn201896m

[97] Tsang MK, Chan CF, Wong KL, Hao J. Comparative studies of upconversion luminescence characteristics and cell bioimaging based on one-step synthesized upconversion nanoparticles capped with different functional groups. Journal of Luminescence. 2015;**157**:172-178. DOI: 10.1016/j.jlumin.2014.08.057

[98] Chávez-García D, Juarez-Moreno K, Reyes R, Barrera J, Hirata GA. Nanotoxicological assessments to warranty the use of functionalized $Y_2O_3$ nanoparticles for biomedical applications. Advanced Materials Letters. 2020;**11**(12):1-9. DOI: 10.5185/amlett.2020.121583

[99] Del Rosal B, Jaque D. Upconversion nanoparticles for *in vivo* applications: Limitations and future perspectives. Methods and Applications in Fluorescence. 2019;7(2):022001. DOI: 10.1088/2050-6120/ab029f

[100] Ortgies DH, Tan M, Ximendes EC, del Rosal B, Hu J, Xu L, et al. Lifetime-encoded infrared-emitting nanoparticles for *in vivo* multiplexed imaging. ACS Nano. 2018;**12**(5):4362-4368. DOI: 10.1021/acsnano.7b09189

[101] Zhan Q, Qian J, Liang H, Somesfalean G, Wang D, He S, et al. Using 915 nm laser excited $Tm^{3+}$/$Er^{3+}$/$Ho^{3+}$–doped NaYbF4 upconversion nanoparticles for *in vitro* and deeper *in vivo* bioimaging without overheating irradiation. ACS Nano.

2011;**5**(5):3744-3757. DOI: 10.1021/nn200110j

[102] Xiong LQ, Chen ZG, Yu MX, Li FY, Liu C, Huang CH. Synthesis, characterization, and *in vivo* targeted imaging of amine-functionalized rare-earth up-converting nanophosphors. Biomaterials. 2009;**30**(29):5592-5600. DOI: 10.1016/j.biomaterials.2009.06.015

[103] Zeng S, Tsang MK, Chan CF, Wong KL, Hao J. PEG modified BaGdF5:Yb/Er nanoprobes for multi-modal upconversion fluorescent, *in vivo* X-ray computed tomography and biomagnetic imaging. Biomaterials. 2012;**33**(36):9232-9238. DOI: 10.1016/j.biomaterials.2012.09.019

[104] Hilderbrand SA, Shao F, Salthouse C, Mahmood U, Weissleder R. Upconverting luminescent nanomaterials: Application to *in vivo* bioimaging. Chemical Communications. 2009;**28**:4188-4190. DOI: 10.1039/B905927J

[105] Han S, Deng R, Xie X, Liu X. Enhancing luminescence in lanthanide-doped upconversion nanoparticles. Angewandte Chemie - International Edition. 2014;**53**:11702-11715

[106] Mosmann T. Rapid colorimetric assay for cellular growth and survival: Application to proliferation and cytotoxicity assays. Journal of Immunological Methods. 1983;**65**(1-2):55-63. DOI: 10.1016/0022-1759(83)90303-4

[107] Shan J, Chen J, Meng J, Collins J, Soboyejo W, Friedberg JS, et al. Biofunctionalization, cytotoxicity, and cell uptake of lanthanide doped hydrophobically ligated $NaYF_4$ upconversion nanophosphors. Journal of Applied Physics. 2008;**104**(9):094308. DOI: 10.1063/1.3008028

[108] Di Meo S, Reed TT, Venditti P, Victor VM. Role of ROS and RNS Sources in Physiological and Pathological Conditions. In: Oxidative Medicine and Cellular Longevity. London UK: Hindawi Publishing Corporation; 2016. DOI: 10.1155/2016/1245049

[109] Regoli F, Giuliani ME. Oxidative pathways of chemical toxicity and oxidative stress biomarkers in marine organisms. Marine Environmental Research. 2014;**93**:106-117. DOI: 10.1016/j.marenvres.2013.07.006

[110] Shapiro HM. Practical Flow Citometry. Hoboken, New Jeresey: John Wiley & Sons, Inc; 2003. p. 724

[111] Van Engeland M, Nieland LJW, Ramaekers FCS, Schutte B, Reutelingsperger CPM. Annexin V-affinity assay: A review on an apoptosis detection system based on phosphatidylserine exposure. Cytometry. 1998;**31**(1): 1-9. DOI: 10.1002/(SICI)1097-0320(19980101)31:1<1::AID-CYTO1>3.0.CO

[112] Riedl SJ, Shi Y. Molecular mechanisms of caspase regulation during apoptosis. Nature Reviews. Molecular Cell Biology. 2004;**5**(11):897-907. DOI: 10.1038/nrm1496

[113] Vermes I, Haanen C, Steffens-Nakken H, Reutellingsperger C. A novel assay for apoptosis flow cytometric detection of phosphatidylserine expression on early apoptotic cells using fluorescein labelled Annexin V. Journal of Immunological Methods. 1995;**184**(1):39-51. DOI: 10.1016/0022-1759(95)00072-I

[114] Das GK, Sudheendra L, Kennedy IM. Heterogeneous nanostructures for plasmonic interaction with luminescence and quantitative surface-enhanced Raman spectroscopy. Reporters, Markers, Dyes, Nanoparticles, and Molecular Probes for Biomedical Applications VI. 2014;**8956**:89560X. DOI: 10.1117/12.2037889

[115] Stone N, Kendall C, Smith J, Crow P, Barr H. Raman spectroscopy for identification of epithelial cancers. Faraday Discussions. 2004;**126**:141-157. DOI: 10.1039/B304992B

[116] Brauchle E, Thude S, Brucker SY, Schenke-Layland K. Cell death stages in single apoptotic and necrotic cells monitored by Raman microspectroscopy. Scientific Reports. 2014;**4**:4698. DOI: 10.1038/srep04698

[117] Karlsson HL, Di Bucchianico S, Collins AR, Dusinska M. Can the comet assay be used reliably to detect nanoparticle-induced genotoxicity? Environmental and Molecular Mutagenesis. 2015;**56**(2):82-96. DOI: 10.1002/em.21933

[118] Azqueta A, Gutzkow KB, Priestley CC, Meier S, Walker JS, Brunborg G, et al. A comparative performance test of standard, medium- and high-throughput comet assays. Toxicology in Vitro. 2013;**27**(2):768-773. DOI: 10.1016/j.tiv.2012.12.006

[119] Speit G, Hartmann A. The comet assay. In: Henderson DS, editor. DNA Repair Protocols: Mammalian Systems. Totowa, NJ: Humana Press; 2006. pp. 275-286. DOI: 10.1385/1-59259-973-7:275

[120] Hamblin MR. Upconversion in photodynamic therapy: Plumbing the depths. Dalton Transactions. 2018;**47**(26):8571-8580. DOI: 10.1039/C8DT00087E

[121] Nigoghossian K, Saska S, Christovam LM, Coelho F, Beatrice CAG, Lucas AA, et al. Upconversion 3D printed composite with multifunctional applications for tissue engineering and photodynamic therapy. Journal of the Brazilian Chemical Society. 2020;**31**(4):638-652. DOI: 10.21577/0103-5053.20190228

[122] Doronkina A, Kazadaeva N, Sagaidachnaya E, Yanina I, Skaptsov A, Konyukhova Y, et al. Erythrocyte aggregation stimulated by NaYF4: $Er^{3+}$, $Yb^{3+}$ upconversion nanoparticles. Proceedings Saratov Fall Meeting 2019: Optical and Nano-Technologies for Biology and Medicine. 2020;**114570**:11457D. DOI: 10.1117/12.2559853

[123] Chatterjee DK, Rufaihah AJ, Zhang Y. Upconversion fluorescence imaging of cells and small animals using lanthanide doped nanocrystals. Biomaterials. 2008;**29**(7):937-943. DOI: 10.1016/j.biomaterials.2007.10.051

[124] Wang M, Mi CC, Wang WX, Liu CH, Wu YF, Xu ZR, et al. Immunolabeling and NIR-excited fluorescent imaging of HeLa cells by using NaYF4: Yb, Er upconversion nanoparticles. ACS Nano. 2009;**3**(6):1580-1586. DOI: 10.1021/nn900491j

[125] Abdul Jalil R, Zhang Y. Biocompatibility of silica coated NaYF4 upconversion fluorescent nanocrystals. Biomaterials. 2008;**29**(30):4122-4128. DOI: 10.1016/j.biomaterials.2008.07.012

[126] Punjabi A, Wu X, Tokatli-Apollon A, El-Rifai M, Lee H, Zhang Y, et al. Amplifying the red-emission of upconverting nanoparticles for biocompatible clinically used prodrug-induced photodynamic therapy. ACS Nano. 2014;**8**(10):10621-10630. DOI: 10.1021/nn505051d

[127] Jiang S, Zhang Y, Lim KM, Sim EKW, Ye L. NIR-to-visible upconversion nanoparticles for fluorescent labeling and targeted delivery of siRNA. Nanotechnology. 2009;**20**(15):155101. DOI: 10.1088/0957-4484/20/15/155101

[128] Vetrone F, Naccache R, Juarranz De La Fuente A, Sanz-Rodríguez F, Blazquez-Castro A, Rodriguez EM, et al.

Intracellular imaging of HeLa cells by non-functionalized NaYF 4: $Er^{3+}$, $Yb^{3+}$ upconverting nanoparticles. Nanoscale. 2010;**2**(4):495-498. DOI: 10.1039/B9NR00236G

[129] Hu H, Yu M, Li F, Chen Z, Gao X, Xiong L, et al. Facile epoxidation strategy for producing amphiphilic up-converting rare-earth nanophosphors as biological labels. Chemistry of Materials. 2008;**20**(22):7003-7009. DOI: 10.1021/cm801215t

[130] Juarez-Moreno K, Chávez-García D, Hirata G, Vazquez-Duhalt R. Monolayer (2D) or spheroids (3D) cell cultures for nanotoxicological studies. Comparison of cytotoxicity and cell internalization of nanoparticles. Toxicology in Vitro. 2022;**85**:105461. DOI: 10.1016/j.tiv.2022.105461

[131] Liu Y, Hou W, Sun H, Cui C, Zhang L, Jiang Y, et al. Thiol-ene click chemistry: A biocompatible way for orthogonal bioconjugation of colloidal nanoparticles. Chemical Science. 2017;**8**(9):6182-6187. DOI: 10.1021/cm801215t

[132] Chen Y, D'Amario C, Gee A, Duong HTT, Shimoni O, Valenzuela SM. Dispersion stability and biocompatibility of four ligand-exchanged NaYF4: Yb, Er upconversion nanoparticles. Acta Biomater. 2020;**102**:384-393. DOI: 10.1016/j.actbio.2019.11.048

[133] M. Ekkapongpisit, A. Giovia, C. Follo, G. Caputo & . Isidoro, C. Biocompatibility, endocytosis, and intracellular trafficking of mesoporous silica and polystyrene nanoparticles in ovarian cancer cells: Effects of size and surface charge groups. International Journal of Nanomedicine 2012; 2012. 4147-4158. DOI: 10.2147/IJN.S33803

[134] Ichimura T, Jin T, Fujita H, Higuchi H, Watanabe TM. Nano-scale measurement of biomolecules by optical microscopy and semiconductor nanoparticles. Frontiers in Physiology. 2014;**5**:273. DOI: 10.3389/fphys.2014.00273

[135] Tsai S, Chen Y, Liaw JW. Compound cellular imaging of laser scanning confocal microscopy by using gold nanoparticles and dyes. Sensors. 2008;**8**(4):2306-2316. DOI: 10.3390/s8042306

Chapter 6

# The Future Applications for Luminescent Material in Optoelectronics and Wearable Devices

*Reza Zarei Moghadam and Amir Hossein Farahani*


## Abstract

Optoelectronic devices, such as LEDs (light-emitting diodes) and OLEDs (organic light-emitting diodes), have a promising future with luminescent materials. These materials play a crucial role in enhancing device performance, efficiency, and functionality. Advances in luminescent materials, including quantum dots, perovskites, and organic dyes, are driving innovations in displays, lighting, sensing, and communication technologies. The future holds potential for even more efficient and versatile optoelectronic devices with the continued development of novel luminescent materials and fabrication techniques. Flexible and wearable devices are one of the future usages for luminescent materials compatible with flexible substrates. Future research may focus on enhancing the durability, stretchability, and comfort of these devices, opening up new opportunities in wearable technology, smart textiles, and flexible displays. This could involve optimizing the spectral characteristics, stability, and energy efficiency of luminescent materials to meet the demanding requirements of wearable devices.


**Keywords:** optoelectronic devices, light-emitting diodes, luminescent materials, quantum dots, wearable devices

## 1. Introduction

Optoelectronic devices that harness light energy have revolutionized various fields, from telecommunications to sensing and imaging [1]. The advent of luminescent materials, particularly two-dimensional (2D) materials, has opened new avenues for optoelectronics, LEDs, infrared photodetectors, and wearable photovoltaics [2]. Spectroscopy is closely associated with luminescence, which is the examination of matter's general laws that govern radiation absorption and emission [3]. The three primary modes of luminescence are fluorescence, phosphorescence and chemiluminescence. Photoluminescence can be classified into two types: fluorescence and



IntechOpen

phosphorescence. In photoluminescence, the glow of a substance is caused by light, while in chemiluminescence, the glow is caused by a chemical reaction. Fluorescence and phosphorescence are dependent on the absorption and emission of light by substances, with longer wavelengths at lower energy levels and time is the most significant divergence [4–6]. Fluorescence results in an immediate emission, which is typically detectable only if the light source is continuously powered on. On the other hand, phosphorescent materials can store the absorbed light energy for a while and then re-emit the light, resulting in a persistent afterglow even after the light has disappeared [7]. The mechanism for relaxing the excited state is similar to that of fluorescence, but chemiluminescence is characterized by the unique generation of excited states. Unlike exothermic reactions, which release energy in the form of heat, certain chemical reactions produce electronically excited products. When the excited state is relaxed by luminescence due to the emission of photons, it is called chemiluminescence. When a biological enzyme catalyzes a reaction, it is called bioluminescence, even though the mechanism is the same [8].

## 2. The role of luminescent materials in optoelectronics

In the realm of optoelectronics, luminescent materials play a crucial role in the generation, manipulation, and detection of light. They are used to fabricate light-emitting diodes (LEDs), lasers, and photodetectors. Luminescent materials used in LEDs emit light of a specific wavelength when an electric current is passed through them. This property allows for the production of highly efficient and long-lasting lighting sources. In addition, luminescent materials are employed in lasers to achieve precise control over the emitted light, enabling applications in telecommunications, medical treatments, and scientific research [9]. Moreover, luminescent materials find application in photodetectors, which convert light into electrical signals. These devices are essential in various industries, including aerospace, automotive, and security systems. Luminescent materials enhance the sensitivity and response time of photodetectors, enabling them to detect light at low levels and rapid speeds. The ability to integrate luminescent materials into optoelectronic devices has revolutionized the field and opened up new avenues for technological advancements [10]. **Figure 1** shows the mechanism of (a) LED with white color and (b) RGB-LED.

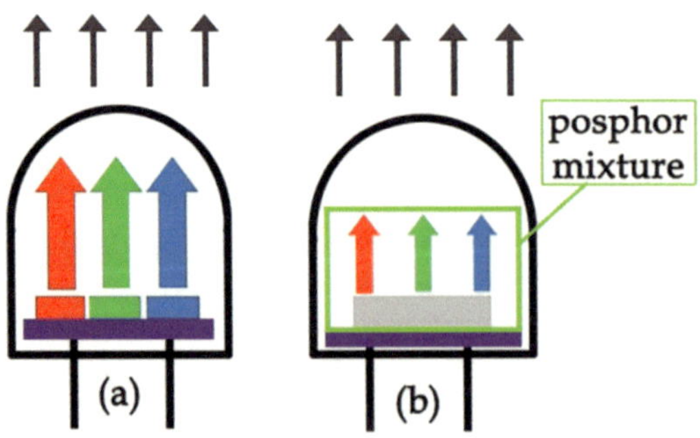


**Figure 1.**
*Mechanism of providing light-emitting diode with white color. (a) Combination of three red-green-blue (RGB) LEDs and (b) mixing of different luminescent materials to generate the white color.*

## 3. Applications of luminescent materials in wearable devices

The integration of luminescent materials into wearable devices has brought about a paradigm shift in the way we interact with technology. Luminescent materials offer unique properties that make them ideal for wearable applications. One such application is in flexible displays. This study demonstrates the great potential of photochromic fibers in wearable human-machine integration and advances the field of wearable human-computer interaction [11], where luminescent materials are used to create vibrant and energy-efficient screens. These displays can be seamlessly integrated into clothing or accessories, providing users with real-time information and enhancing their overall experience and medicine, as like as photomedicine for curing the cancers. The high potential of QLED to accelerate the adoption of photomedicine is expected to broadly cover multiple healthcare markets, including cancer treatment, periodontics, dermatology (especially cosmetic dermatology), and chronic wound and ulcer treatment [12]. The luminescent spectra of the 22 μm B-Si layered sample and the 31 μm B-Si layered sample are displayed in **Figure 2a** and **b**, respectively. Upon observing the figures, it becomes apparent that B-Si exhibits a strong visible emission at room temperature; however, the emitted spectrum shows slight variations. In **Figure 3a**, there are two distinct emission peaks. The first peak represents a sharp UV-blue emission ranging from 380 to 500 nm, while the second peak corresponds to a broader red emission centered on 650 nm. The inset of **Figure 2a** displays a camera image that captures a similar emission pattern. On the other hand, **Figure 2b** only exhibits a single emission peak in its photoluminescence spectra, which corresponds to the UV-blue emission. This observation is further confirmed by the camera image shown in the inset. The number of photoluminescence peaks in a spectrum provides valuable information regarding the diameter or crystal size of nanostructures [13].

Furthermore, luminescent materials are employed in the development of smart textiles [15], which can monitor vital signs, detect environmental conditions, and even generate power. By embedding luminescent materials into the fabric, wearable devices can become more responsive and adaptive to the wearer's needs [16]. One of the most promising aspects of these luminescent materials is their sensitivity to temperature changes [17]. When exposed to different temperatures, the gels exhibit a distinct shift in their emission color. This makes them potentially useful as temperature sensors in various applications. For instance, luminescent materials can be used

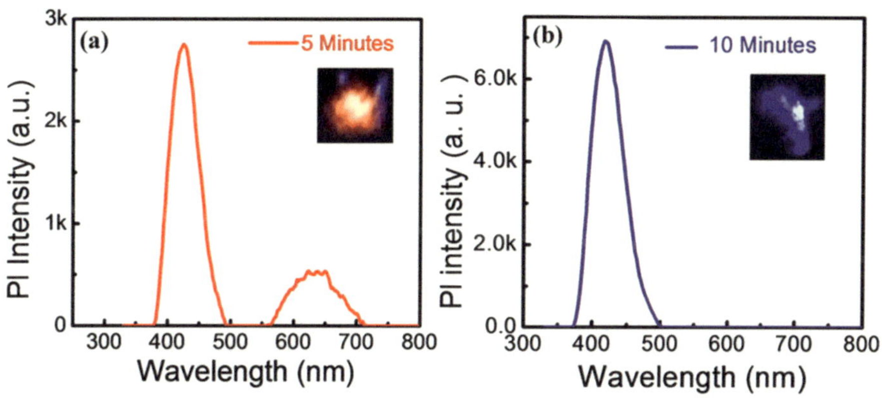


**Figure 2.**
*Luminescence spectra of transferred B-Si NSs on flexible substrate [13].*

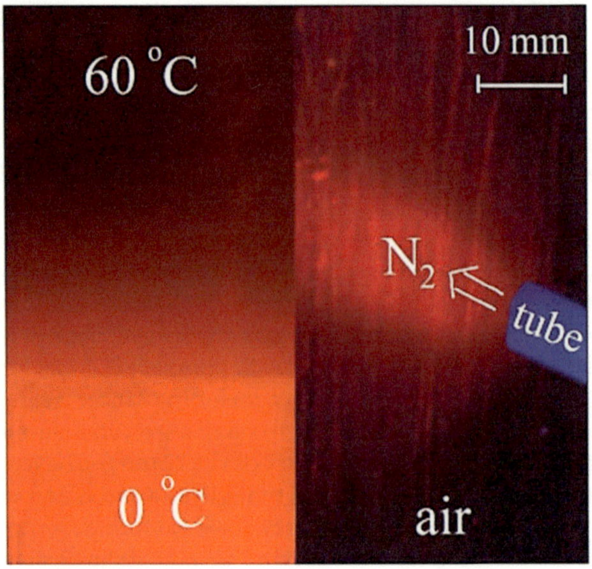


**Figure 3.**
*Luminescence images of the sensor foil. Left: temperature gradient imaged through a Chroma 580 bandpass filter. Right: oxygen partial pressure for the same sensor foil imaged through an RG 650 nm long pass filter at an excitation wavelength of 366 nm [14].*

to create sensors that change color in response to temperature or chemical changes, alerting the wearer to potential dangers or providing valuable feedback [17]. **Figure 3** represents the luminescence images of the sensor foil where the left picture shows temperature gradient imaged through a Chroma 580 bandpass filter and right picture shows oxygen partial pressure for the same sensor foil imaged through an RG 650 nm long pass filter at an excitation wavelength of 366 nm [14].

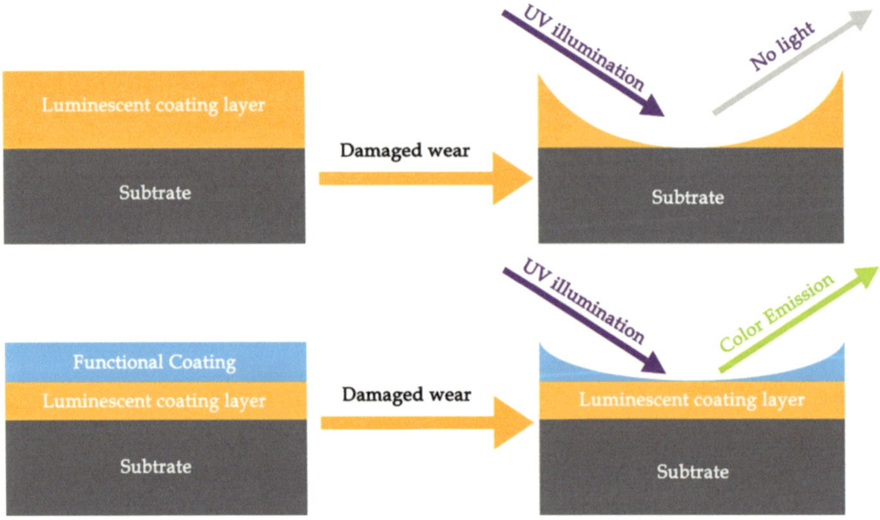


**Figure 4.**
*Schematic diagrams showing luminescent particles code posited as top layer in (a) or interlayer in (b), and the coating appearance in the damaged area.*

The unique properties of these luminescent materials make them ideal candidates for a variety of sensing applications. For instance, they could be used as coatings to monitor the structural integrity of pipes, cables, and other underwater structures critical to offshore energy operations. In addition, they could be used to detect chemical variations in liquids or to observe velocity gradients in fluid flow experiments [18, 19]. In the survey that Y. He1, S.C. Wang, and their colleagues have done before, it shows the luminescent material does not scatter any light in visible width waves, but it emits in UV light. For the coating in **Figure 4a**, the vanishing of luminescence under UV light is a sign of coating damage and may require recoating. By comparison, the attention of colorful light in **Figure 4b** suggests that the active coating is worn away [14].

## 4. Advancements in luminescent materials for optoelectronics and wearable devices

The field of luminescent materials is constantly evolving, with researchers and industry players striving to improve their properties and performance. Recent advancements have focused on enhancing the efficiency, stability, and tunability of luminescent materials. One such breakthrough is the development of perovskite-based luminescent materials as pointed in previous sections, which have shown exceptional performance in optoelectronic devices. These materials offer high quantum efficiency [20], narrow emission peaks [21], and excellent long-term stability [22], making them promising candidates for future applications.

Additionally, the integration of luminescent materials with nanotechnology has opened up new possibilities for optoelectronics and wearable devices [23]. Nanomaterials, such as quantum dots, exhibit unique size-dependent luminescent properties, allowing for precise control over the emitted light. These materials can be incorporated into flexible substrates [24], enabling the development of bendable and stretchable optoelectronic devices. Furthermore, the use of nanomaterials enhances the efficiency and color purity of luminescent materials, making them highly desirable for displays and lighting applications [12]. Luminescent materials can produce with various techniques; some of them are shown in **Table 1**.

| Techniques | Extract from | Reference |
|---|---|---|
| Bioluminescence | Luciferase | [25] |
| Chemiluminescence | Enthalpy of the reactions | [26] |
| Crystalloluminescence | Formation of bonds between the atoms | [27] |
| Electroluminescence | Passing of electric current | [28] |
| Cathodoluminescence | Photons | [29] |
| Radioluminescence | Ionizing radiations like $\alpha$, $\beta$, or $\gamma$ rays | [30] |
| Sonoluminescence | Small bubbles in a solvent on excitation by ultrasonic waves | [28] |
| Thermoluminescence | Eating crystalline materials | [31] |

**Table 1.**
*The luminescence producing techniques.*

### 4.1 Circularly polarized luminescence (CPL)

One of the advancements in luminescent materials for optoelectronics is circularly polarized luminescence (CPL). Traditionally, circularly polarized light can be generated from unpolarized light through the physical method (**Figure 5(a)**). The emitted unpolarized light is first converted into linearly polarized light by the linear polarizer, and then further decomposed into left or right circularly polarized light through the quarter-wave plate. During this indirect physical process, at least 50% of energy will be lost [14]. Therefore, it is urgent to develop novel luminescent materials that can directly generate circularly polarized light. But in a new method referred to the geometric property of an object and Chiral luminophore, the direct polarized light can be produced with decreasing the lost energy as shown in **Figure 5(b)** [32].

### 4.2 Perovskite quantum dots (PQDs)

Perovskite-based materials have emerged as promising candidates for optoelectronic applications due to their high photoluminescence quantum yield, tunable bandgap, and low-cost fabrication. PQDs offer excellent color purity and high brightness, making them suitable for displays and lighting [33]. Most studies have focused on colloidal quantum dots (QDs) due to their high photoluminescence (PLQY), tunable wavelength, and thin emission wavelength. Using the quantum confinement effect, the emission color of quantum dots can be controlled depending on the size and content of the quantum dots. These advantages make QDs useful in solar cells, lasers, light-emitting diodes (LEDs), and bioimaging [33].

### 4.3 Flexible and stretchable luminescent materials

The development of flexible and stretchable luminescent materials is crucial for wearable optoelectronic devices. Recent progress in materials engineering, including stretchable polymers, elastomers, and nanocomposites, has enabled the fabrication of wearable displays, health monitors, and smart textiles with conformal and

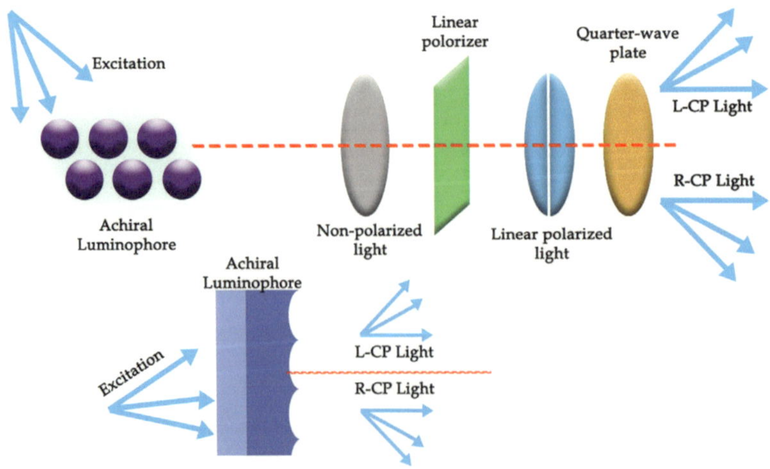


**Figure 5.**
*Two techniques for producing circularly polarized light. (a) Traditional method and (b) circularly polarized luminescence.*

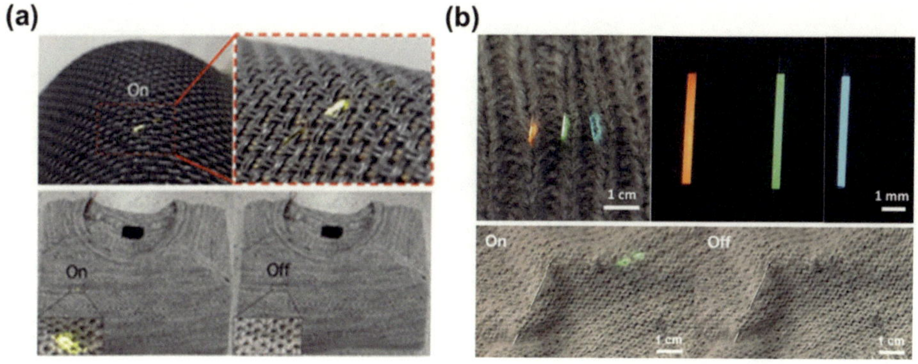


**Figure 6.**
*(a) Fibrous OLEDs handwoven into clothing and (b) fibrous multicolored phOLEDs woven into everyday clothing [35].*

stretchable light-emitting components [34]. Wearable displays that can adapt to the contours of the human body are of great interest for real-time visual expression and communication [35]. **Figure 6a** shows an electronic textile for wearable displays with DC-powered LEDs with luminous efficiency that you can compare with traditional OLEDs on glass [36]. On the other image, **Figure 6b** shows an electronic textile for wearable displays with phOLED, which has higher internal efficiency and lower driving voltage compared to OLED [37].

### 4.4 Bioinspired luminescent materials

Inspired by natural photonic structures, researchers have developed biomimetic luminescent materials with enhanced light extraction efficiency and color-tuning capabilities. These materials, often based on photonic crystals, structural colors, and animals (insects)

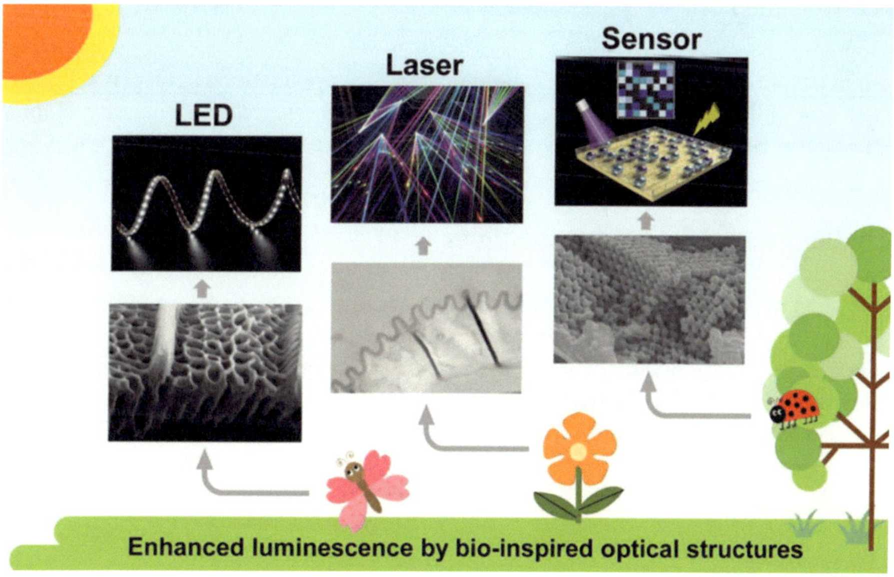


**Figure 7.**
*Image of optical structures and applications to LEDs, lasers, and sensors to improve the luminescence observed by living organisms [24].*

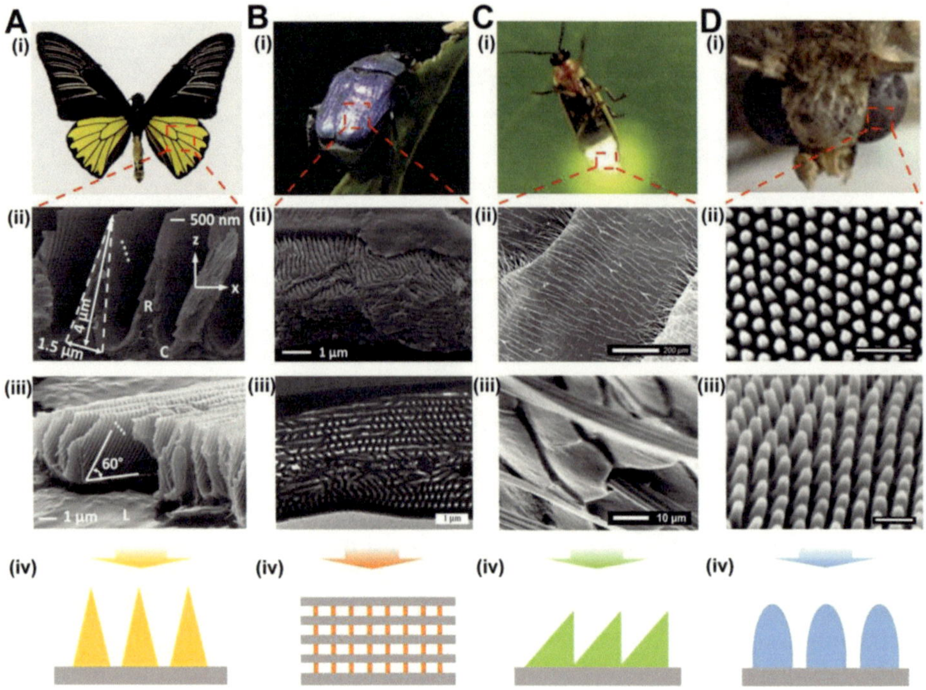


**Figure 8.**
*The hind wings of the Troïdes magellanus butterfly exhibit a uniform yellow color in sunlight and an increasing yellow-green color under ultraviolet illumination. Scanning electron microscopy (SEM) cross-sectional and side-view images show that the ridge structures with triangular cross sections are set as a lattice structure, with each ridge having a series of lamellae [39]. (B) Hoplia Correa beetle exhibits a purple-blue color. The scales covering the elytra are composed of alternating membranes of pure and mixed porous air cuticle layers, forming a periodic photonic structure [40]. (C) Fireflies can release bright light from their abdomens, which have complicated optical structures. Ventral scale mismatch improves emission extraction [41]. (D) The corneal surface of a moth eye is lined with nano pill structures—reflective construction increases light input [24, 42].*

found in nature as shown in **Figure 7**, hold promise for applications in displays, lighting, and sensing [38]. The structures of some insects as *Troïdes magellanus* butterfly, Hoplia Correa beetle, fireflies, and moth eye are shown in **Figure 8** with their SEM images [24].

## 5. Challenges and future perspectives

The cost of luminescent materials can be a barrier to their implementation in large-scale applications. Some luminescent materials, such as rare-earth elements, are expensive and difficult to source. This restricts their use to niche markets and high-end devices. To drive widespread adoption, efforts are underway to develop cost-effective alternatives and scalable manufacturing processes (**Figure 9**) [43].

### 5.1 Challenges

#### *5.1.1 Efficiency enhancement*

Efficiency in luminescent materials refers to the ability to convert energy into light effectively. Advances in materials design and engineering can improve this by

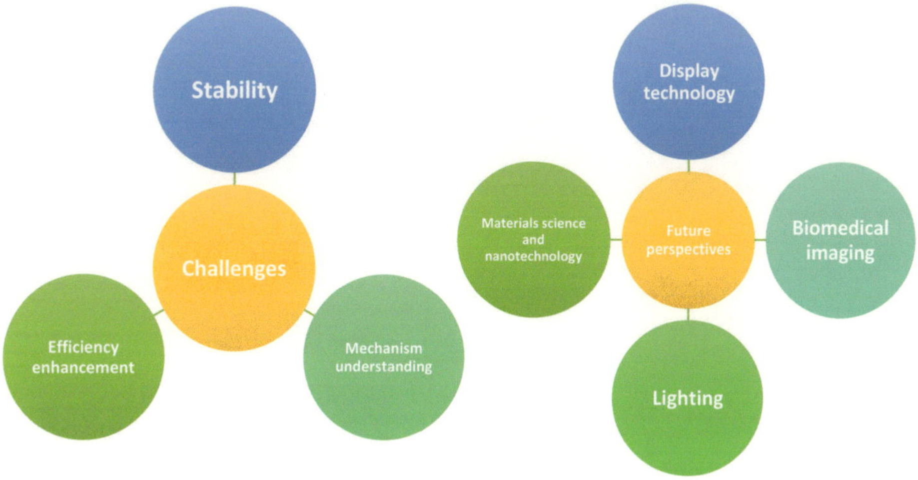


**Figure 9.**
*The challenges and future perspectives of luminescence materials.*

optimizing factors such as quantum yield, lifetime, and energy transfer processes. For instance, incorporating efficient energy transfer mechanisms or minimizing non-radiative decay pathways can enhance overall efficiency. This is crucial for applications like solid-state lighting and displays where energy consumption and performance are key considerations [44].

#### *5.1.2 Stability*

Stability is essential for the long-term performance and reliability of luminescent materials, especially in practical applications where exposure to various environmental factors is inevitable. Research focuses on developing materials with robust chemical and physical properties to withstand degradation over time. Strategies include designing materials with high chemical and thermal stability, encapsulating sensitive components, and exploring novel protection mechanisms to prevent degradation [45]. Some of the stable luminescence are PTM [45], $K_2BaCa(PO_4)_2$ [46], and so on as shown in **Figure 10**.

#### *5.1.3 Mechanism understanding*

Understanding the fundamental mechanisms behind luminescence is crucial for optimizing material design and performance. This involves studying the processes involved in light emission, such as electron transitions, energy transfer mechanisms, and material interactions. Advanced spectroscopic techniques and theoretical modeling are used to elucidate these mechanisms, providing insights into how to control and manipulate luminescent properties at the molecular and nanoscale levels [47].

### **5.2 Future perspectives**

#### *5.2.1 Display technology*

Luminescent materials play a key role in display technology, offering advantages such as high brightness, wide color gamut, and energy efficiency. Organic light-emitting diodes (OLEDs) and quantum dot-based displays are prominent examples.

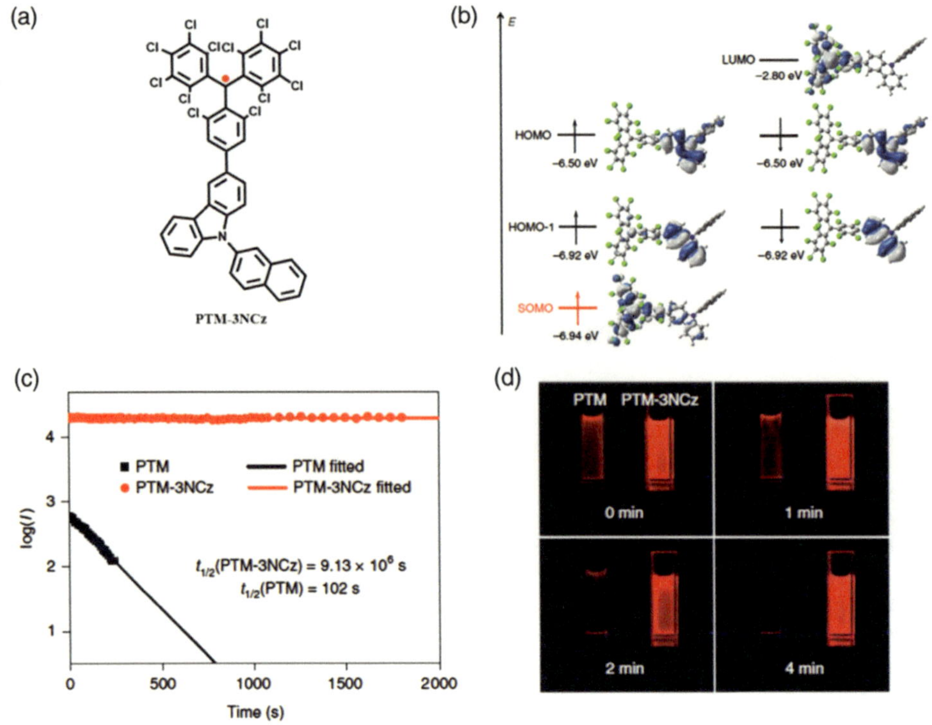


**Figure 10.**
*Optical properties of PTM-3NCz. (a) Chemical structure. (b) Boundary molecular orbital. (c) Comparison of photostability between PTM and PTM-3NCz. (d) Photograph of PTM and PTM-3NCz in dilute cyclohexane solution under UV light [45].*

Ongoing research focuses on improving device efficiency, lifespan, and manufacturability, as well as exploring new display concepts such as flexible and transparent displays for emerging applications in wearables, automotive, and augmented reality [48]. An example of this application is given in **Figure 11**.

### *5.2.2 Lighting*

Luminescent materials are revolutionizing the lighting industry by enabling energy-efficient and environmentally friendly lighting solutions. Solid-state lighting technologies, such as LEDs and phosphor-converted LEDs, utilize luminescent materials to generate light with high efficiency and color quality. Research aims to further enhance efficiency, color rendering, and spectral control while reducing costs and environmental impact, driving the adoption of these technologies in general lighting applications [49].

### *5.2.3 Biomedical imaging*

In biomedical imaging, luminescent probes offer advantages such as high sensitivity, multiplexing capability, and noninvasiveness, making them valuable tools for diagnostics and research. Fluorescent dyes, quantum dots, and upconversion nanoparticles are commonly used for applications such as fluorescence microscopy

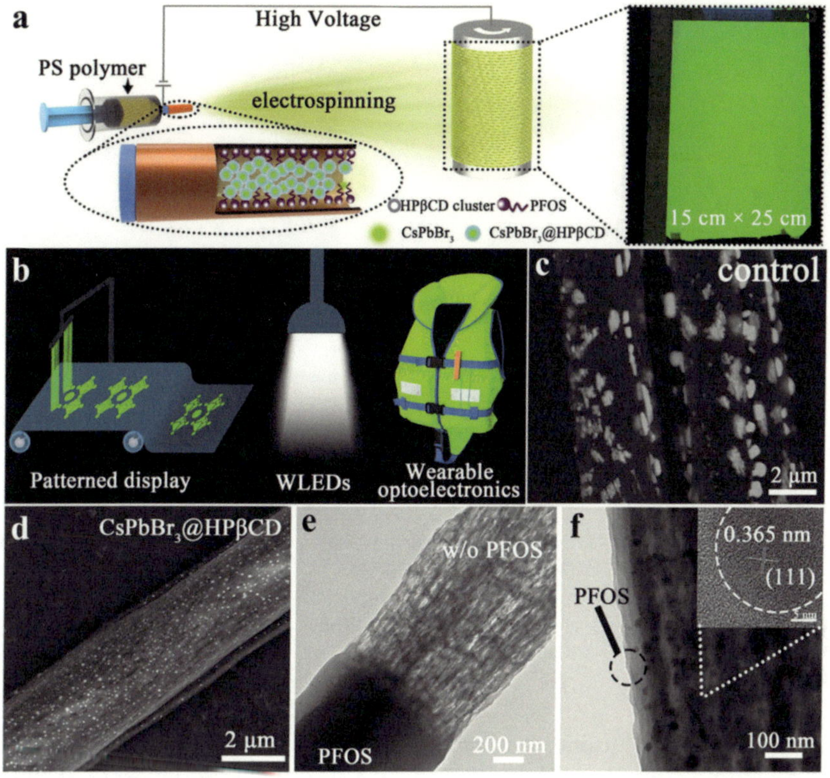


**Figure 11.**
*(a) Schematic illustration of the fabrication process of PLTs and the as-prepared fibrous membrane under UV light irradiation. (b) Application of CsPbBr3@HPβCD@PFOS composites in patterned display, white light-emitting diodes (WLEDs) and wearable optoelectronics. SEM (YAG back-scattered electron detector) images of the (c) control fiber and (d) CsPbBr3@HPβCD fiber. (e) TEM image showing the CsPbBr3@HPβCD fiber without or with PFOS coating as indicated. Note: The top half of the CsPbBr3@HPβCD@PFOS fiber was selectively etched by n-hexane to expose the inner CsPbBr3@HPβCD. (f) Close-up TEM image of the CsPbBr3@HPβCD@PFOS fiber. Inset: HRTEM image of the CsPbBr3 crystal.*

(as shown in **Figure 12(a)**), molecular imaging, and drug delivery tracking. Ongoing research focuses on improving probe specificity, biocompatibility, and imaging depth for applications in disease diagnosis, drug development, and personalized medicine [51]. **Figure 12(b)** shows a cell that has been pictured by wide-field fluorescence microscope.

### *5.2.4 Materials science and nanotechnology*

Materials science and nanotechnology play pivotal roles in advancing luminescent materials and their applications. Researchers explore novel synthesis methods, nanostructuring techniques, and hybrid material systems to tailor luminescent properties for specific applications. Key areas of focus include improving material stability, enhancing quantum efficiency, and achieving precise control over optical properties. These advancements drive innovation across various fields, from electronics and photonics to healthcare and energy, opening up new opportunities for luminescence-based technologies [46].

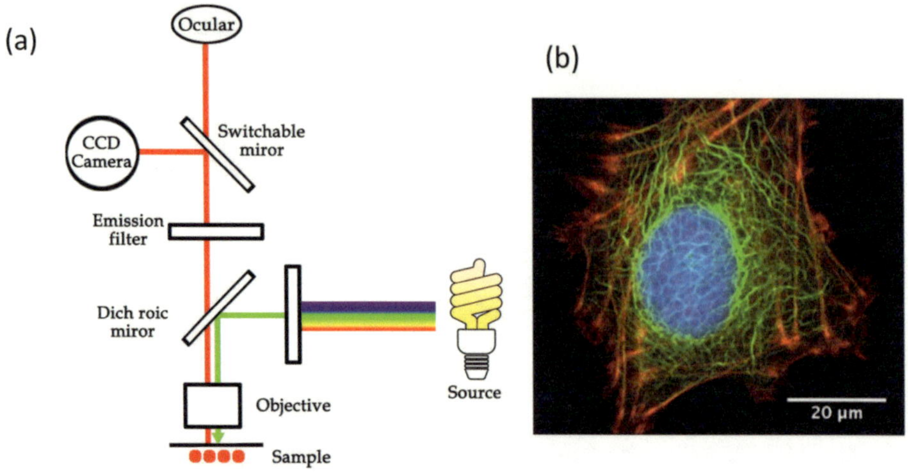


**Figure 12.**
*(a) Anatomical representation of the wide-field fluorescence microscope setup. (b) Fluorescence microscope image of a cell, where the mesh of keratin average filaments is shown in green, actin filaments in red, and nuclear DNA in blue [50].*

## 6. Conclusion

The applications of luminescent materials in optoelectronics and wearable devices are vast and promising. From leveraging the unique optical properties of 2D materials in LEDs and photodetectors to harnessing wavelength conversion and emission properties for sensing and imaging, these materials offer exciting opportunities. However, challenges remain, including improving quantum efficiency prediction, developing environmentally benign materials, and enhancing material lifetime and scalability. As the demand for energy-efficient lighting, innovative sensing functions, and advanced optoelectronic devices continues to grow, the luminescent materials market is poised for substantial expansion. Overcoming the hurdles through data-driven approaches, composition and nanoarchitecture engineering, and photonic engineering will be crucial in unlocking the full potential of these materials in next-generation technologies.

## Acknowledgments

The authors have used Grammarly software to edit the language.

## Author details

Reza Zarei Moghadam* and Amir Hossein Farahani
Faculty of Science, Physics Department, Arak University, Arak, Iran

*Address all correspondence to: r-zareimoghadam@araku.ac.ir

## References


[1] Kumbhakar P, Chowde Gowda C, Tiwary CS. Advance optical properties and emerging applications of 2D materials. Frontiers in Materials. 2021;**8**:721514

[2] Jia B. 2D optical materials and the implications for photonics. APL Photonics. 2019;**4**:080401-080405

[3] Chopra V, Kadam AR, Swart HC, Dhoble SJ. Chapter 1—Introduction to luminescence and radiation dosimetry techniques. In: Dhoble S, Chopra V, Nayar V, Kitis G, Poelman D, Swart H, editors. Radiation Dosimetry Phosphors. United Kingdom: Woodhead Publishing; 2022. pp. 1-27

[4] Lichtman JW, Conchello J-A. Fluorescence microscopy. Nature Methods. 2005;**2**:910-919

[5] Baryshnikov G, Minaev B, Ågren H. Theory and calculation of the phosphorescence phenomenon. Chemical Reviews. 2017;**117**:6500-6537

[6] Dodeigne C, Thunus L, Lejeune R. Chemiluminescence as diagnostic tool. A review. Talanta. 2000;**51**:415-439

[7] Ni A-Y, Zhao H, Zhang P-P, Zhang B-L, Zhang J-J, Liu S, et al. A reversible room temperature phosphorescence/delayed fluorescence switch trigged by solvent exchange in a Ca-based coordination polymer. Inorganic Chemistry Frontiers. 2023;**10**:4065-4075

[8] Yang M, Huang J, Fan J, Du J, Pu K, Peng X. Chemiluminescence for bioimaging and therapeutics: Recent advances and challenges. Chemical Society Reviews. 2020;**49**:6800-6815

[9] Li Q, Li Z. Molecular packing: Another key point for the performance of organic and polymeric optoelectronic materials. Accounts of Chemical Research. 2020;**53**:962-973

[10] Moghadam RZ, Ahmadvand H, Jannesari M. Design and fabrication of multi-layers infrared antireflection coating consisting of ZnS and Ge on ZnS substrate. Infrared Physics and Technology. 2016;**75**:18-21

[11] Li Y, Su Y. Mass-produced and uniformly luminescent photochromic fibers toward future interactive wearable displays. Light: Science and Applications. 2024;**13**:79

[12] Triana MA, Hsiang E-L, Zhang C, Dong Y, Wu S-T. Luminescent nanomaterials for energy-efficient display and healthcare. ACS Energy Letters. 2022;**7**:1001-1020

[13] Kumar A, Agrawal J, Sharma AK, Singh V, Agarwal A. A cost-effective and facile approach for realization of black silicon nanostructures on flexible substrate. Journal of Materials Science: Materials in Electronics. 2019;**30**:16554-16561

[14] Borisov SM, Vasylevska AS, Krause C, Wolfbeis OS. Composite luminescent material for dual sensing of oxygen and temperature. Advanced Functional Materials. 2006;**16**:1536-1542

[15] Tian T, Yang M, Fang Y, Zhang S, Chen Y, Wang L, et al. Large-area waterproof and durable perovskite luminescent textiles. Nature Communications. 2023;**14**:234

[16] Zhao K, Zhao Y, Qian R, Ye C, Song Y. Recent advances in interactive mechanosensory electronics with luminescence/coloration outputs for

wearable applications. ACS Materials Letters. 2023;**5**:3093-3116

[17] Zhao Y, Wang X, Zhang Y, Li Y, Yao X. Optical temperature sensing of up-conversion luminescent materials: Fundamentals and progress. Journal of Alloys and Compounds. 2020;**817**:152691

[18] Xu S, Chang L, Liu B, Zhao H, Chu Z, Gong Y, et al. Characterization of image intensifier tubes in vary temperature environments. In: Eighth Symposium on Novel Photoelectronic Detection Technology and Applications. SPIE; 2022;**12169**:3209-3213

[19] Sun K, Cui W, Chen C. Review of underwater sensing technologies and applications. Sensors. 2021;**21**:7849

[20] Fang T, Wang T, Li X, Dong Y, Bai S, Song J. Perovskite QLED with an external quantum efficiency of over 21% by modulating electronic transport. Science Bulletin. 2021;**66**:36-43

[21] Zhu L, Cao H, Xue C, Zhang H, Qin M, Wang J, et al. Unveiling the additive-assisted oriented growth of perovskite crystallite for high performance light-emitting diodes. Nature Communications. 2021;**12**:5081

[22] Zhang S, Liu Z, Zhang W, Jiang Z, Chen W, Chen R, et al. Barrier designs in perovskite solar cells for long-term stability. Advanced Energy Materials. 2020;**10**:2001610

[23] Yu J, Luo M, Lv Z, Huang S, Hsu H-H, Kuo C-C, et al. Recent advances in optical and optoelectronic data storage based on luminescent nanomaterials. Nanoscale. 2020;**12**:23391-23423

[24] Zhang Z, Vogelbacher F, Song Y, Tian Y, Li M. Bio-inspired optical structures for enhancing luminescence. Exploration. 2023;**3**:20220052

[25] Syed AJ, Anderson JC. Applications of bioluminescence in biotechnology and beyond. Chemical Society Reviews. 2021;**50**:5668-5705

[26] Miao W. Electrogenerated chemiluminescence and its biorelated applications. Chemical Reviews. 2008;**108**:2506-2553

[27] Safonov GP, Shlyapintokh VY, Entelis SG. Crystalloluminescence of organic compounds. Nature. 1965;**205**:1203-1204

[28] Singh D, Tanwar V, Bhagwan S, Singh I. Recent advancements in luminescent materials and their potential applications. Advanced Magnetic and Optical Materials. Sweden: Wiley Online Library; 2016:317-352

[29] Smith DJ, McCartney MR. Microscopy applications|semiconductors. In: Worsfold P, Townshend A, Poole C, editors. Encyclopedia of Analytical Science. 2nd ed. Oxford: Elsevier; 2005. pp. 84-91

[30] Klein JS, Sun C, Pratx G. Radioluminescence in biomedicine: Physics, applications, and models. Physics in Medicine and Biology. 2019;**64**:04TR01

[31] Bos AJ. Theory of thermoluminescence. Radiation Measurements. 2006;**41**:S45-S56

[32] Deng Y, Wang M, Zhuang Y, Liu S, Huang W, Zhao Q. Circularly polarized luminescence from organic micro−/nano-structures. Light: Science and Applications. 2021;**10**:76

[33] Wang HC, Bao Z, Tsai HY, Tang AC, Liu RS. Perovskite quantum dots and their application in light-emitting diodes. Small. 2018;**14**:1702433

[34] Choi HW, Shin D-W, Yang J, Lee S, Figueiredo C, Sinopoli S, et al. Smart

textile lighting/display system with multifunctional fibre devices for large scale smart home and IoT applications. Nature Communications. 2022;**13**:814

[35] Cho S, Chang T, Yu T, Lee CH. Smart electronic textiles for wearable sensing and display. Biosensors. 2022;**12**:222

[36] Kwon S, Kim H, Choi S, Jeong EG, Kim D, Lee S, et al. Weavable and highly efficient organic light-emitting fibers for wearable electronics: A scalable, low-temperature process. Nano Letters. 2018;**18**:347-356

[37] Hwang YH, Kwon S, Shin JB, Kim H, Son YH, Lee HS, et al. Bright-multicolor, highly efficient, and addressable phosphorescent organic light-emitting fibers: Toward wearable textile information displays. Advanced Functional Materials. 2021;**31**:2009336

[38] Qu C, Xu Y, Xiao Y, Zhang S, Liu H, Song G. Multifunctional displays and sensing platforms for the future: A review on flexible alternating current electroluminescence devices. ACS Applied Electronic Materials. 2021;**3**:5188-5210

[39] Van Hooijdonk E, Barthou C, Vigneron JP, Berthier S. Angular dependence of structural fluorescent emission from the scales of the male butterfly *Troïdes magellanus* (Papilionidae). Journal of the Optical Society of America B. 2012;**29**:1104-1111

[40] Mouchet SR, Lobet M, Kolaric B, Kaczmarek AM, Van Deun R, Vukusic P, et al. Controlled fluorescence in a beetle's photonic structure and its sensitivity to environmentally induced changes. Proceedings of the Royal Society B: Biological Sciences. 2016;**283**:20162334

[41] Bay A, Cloetens P, Suhonen H, Vigneron JP. Improved light extraction in the bioluminescent lantern of a *Photuris firefly* (Lampyridae). Optics Express. 2013;**21**:764-780

[42] Oh YJ, Kim JJ, Jeong KH. Biologically inspired biophotonic surfaces with self-antireflection. Small. 2014;**10**:2558-2563

[43] Crawford SE, Ohodnicki PR, Baltrus JP. Materials for the photoluminescent sensing of rare earth elements: Challenges and opportunities. Journal of Materials Chemistry C. 2020;**8**:7975-8006

[44] Kitai AH. Solid State Luminescence: Theory, Materials and Devices. London: Springer Science and Business Media; 2012

[45] Cui Z, Abdurahman A, Ai X, Li F. Stable luminescent radicals and radical-based LEDs with doublet emission. CCS Chemistry. 2020;**2**:1129-1145

[46] Sharma M, Gungor K, Yeltik A, Olutas M, Guzelturk B, Kelestemur Y, et al. Near-unity emitting copper-doped colloidal semiconductor quantum wells for luminescent solar concentrators. Advanced Materials. 2017;**29**:1700821

[47] Wang Y, Zhao Z, Yuan WZ. Intrinsic luminescence from nonaromatic biomolecules. ChemPlusChem. 2020;**85**:1065-1080

[48] Ha JM, Hur SH, Pathak A, Jeong J-E, Woo HY. Recent advances in organic luminescent materials with narrowband emission. NPG Asia Materials. 2021;**13**:53

[49] Jüstel T, Nikol H, Ronda C. New developments in the field of luminescent materials for lighting and displays. Angewandte Chemie International Edition. 1998;**37**:3084-3103

[50] Manoel B. Scanning X-Ray Nano-Diffraction on Eukaryotic Cells: From Freeze-Dried to Living Cells. Germany: Göttingen University Press; 2014

[51] Wang R, Zhang F. NIR luminescent nanomaterials for biomedical imaging. Journal of Materials Chemistry B. 2014;**2**:2422-2443